THE BRIDGE BETWEEN WORLDS

Also by Gavin Francis

Free for All: Why the NHS Is Worth Saving
Sir Thomas Browne: The Opium of Time
Recovery: The Lost Art of Convalescence
Intensive Care: A GP, A Community & A Pandemic
Island Dreams: Mapping an Obsession
Shapeshifters: On Medicine & Human Change
Adventures in Human Being
Empire Antarctica: Ice, Silence & Emperor Penguins
True North: Travels in Arctic Europe

THE BRIDGE BETWEEN WORLDS

A BRIEF HISTORY OF CONNECTION

GAVIN FRANCIS

CANONGATE

First published in Great Britain in 2024
by Canongate Books Ltd, 14 High Street, Edinburgh EH1 1TE

canongate.co.uk

1

British Library Cataloguing-in-Publication Data
A catalogue record for this book is available on
request from the British Library

ISBN 978 1 80530 013 7

Typeset in Adobe Caslon Pro by
Palimpsest Book Production Ltd, Falkirk, Stirlingshire

Printed and bound by CPI Group (UK) Ltd, Croydon CR0 4YY

in gratitude for
bridge builders & peacemakers,
optimists & visionaries

Bridges are structures carrying roadways, waterways or railways across streams, valleys or other roads or railways, leaving a passageway below.

Encyclopaedia Britannica (1911)

Bridges gone, perhaps the whole world would fall apart.

Margaret Mahy, 'The Bridge Builder'

The perfect diving board, and solution to any problem.

'Bridge', Urban Dictionary

The world is a bridge; pass over it, but build no houses upon it.

The Qur'an

Contents

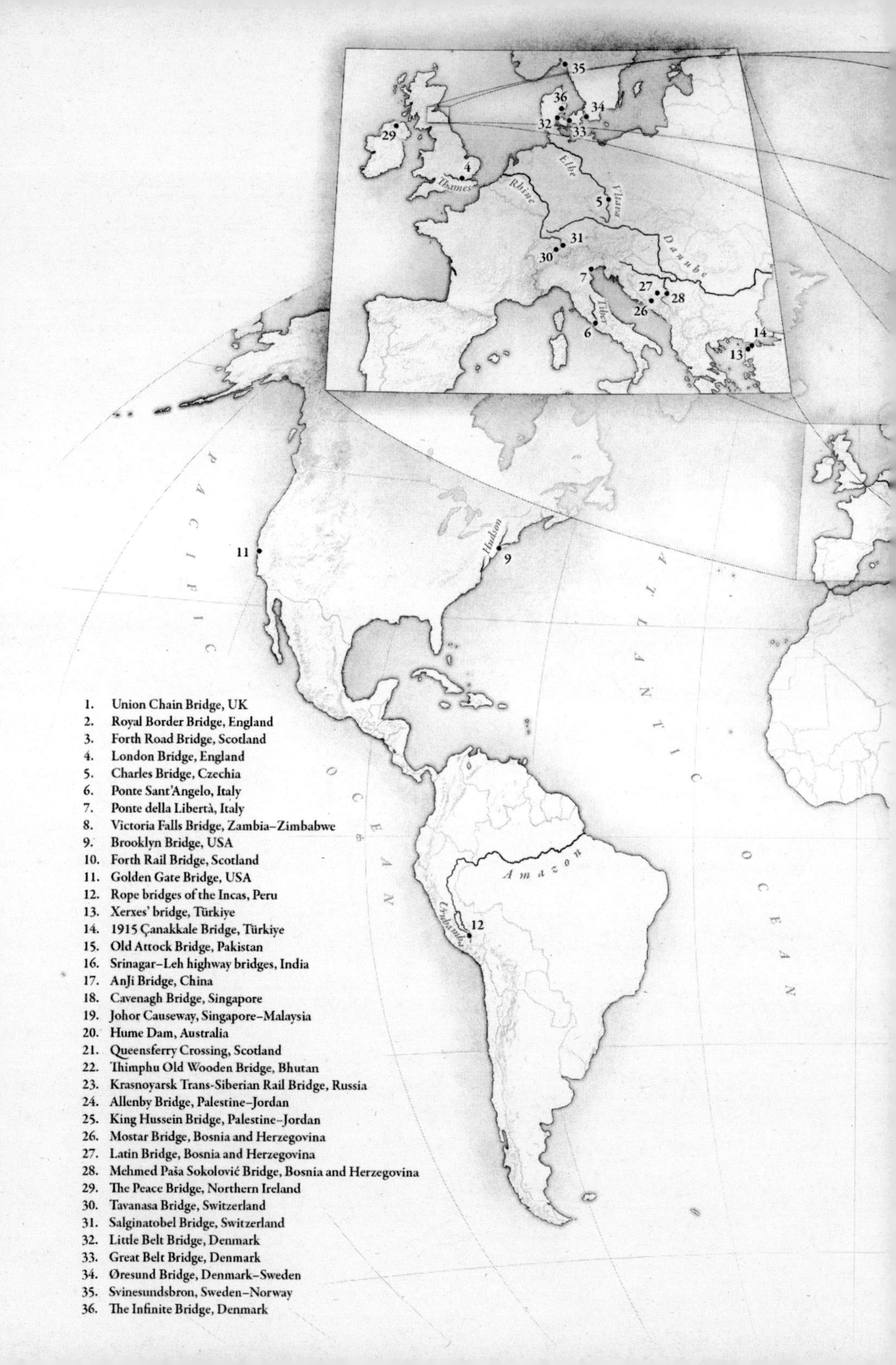
35
36
34
32
33
29
4
Thames
Rhine
Elbe
Vltava
5
31
30
Danube
7
27
28
26
Tiber
6
14
13
P A C I F I C
O C E A N
11
Hudson
9
A T L A N T I C
O C E A N
Amazon
Urubamba
12
1. Union Chain Bridge, UK
2. Royal Border Bridge, England
3. Forth Road Bridge, Scotland
4. London Bridge, England
5. Charles Bridge, Czechia
6. Ponte Sant'Angelo, Italy
7. Ponte della Libertà, Italy
8. Victoria Falls Bridge, Zambia–Zimbabwe
9. Brooklyn Bridge, USA
10. Forth Rail Bridge, Scotland
11. Golden Gate Bridge, USA
12. Rope bridges of the Incas, Peru
13. Xerxes' bridge, Türkiye
14. 1915 Çanakkale Bridge, Türkiye
15. Old Attock Bridge, Pakistan
16. Srinagar–Leh highway bridges, India
17. AnJi Bridge, China
18. Cavenagh Bridge, Singapore
19. Johor Causeway, Singapore–Malaysia
20. Hume Dam, Australia
21. Queensferry Crossing, Scotland
22. Thimphu Old Wooden Bridge, Bhutan
23. Krasnoyarsk Trans-Siberian Rail Bridge, Russia
24. Allenby Bridge, Palestine–Jordan
25. King Hussein Bridge, Palestine–Jordan
26. Mostar Bridge, Bosnia and Herzegovina
27. Latin Bridge, Bosnia and Herzegovina
28. Mehmed Paša Sokolović Bridge, Bosnia and Herzegovina
29. The Peace Bridge, Northern Ireland
30. Tavanasa Bridge, Switzerland
31. Salginatobel Bridge, Switzerland
32. Little Belt Bridge, Denmark
33. Great Belt Bridge, Denmark
34. Øresund Bridge, Denmark–Sweden
35. Svinesundsbron, Sweden–Norway
36. The Infinite Bridge, Denmark

Firth of Forth
Tweed
1
2
Firth of Forth
21
3
10
Yenisei
23
17
15
16
Indus
Jordan
24, 25
22
INDIAN OCEAN
Zambezi
8
19
18
Murray
20

Approach

Bridges are good to think with.

When I was very little, my favourite story was 'The Billy Goats Gruff'. It's an old Norwegian fairy tale: three goats want to cross a bridge to reach a field of sweet grass, but a troll hiding beneath it keeps jumping out to stop them. The pictures in my own copy (Ladybird Books, 1977) enthralled but appalled me: dense jungles of foliage, goats with furious eyes, a pot-bellied troll with tombstone teeth and pointed ears.

The bridge in the story is a barrier, but it is also a kind of test. The two smaller goats cannily persuade the troll to wait for the biggest one and, giving in to greed, it agrees. When the big goat arrives at the bridge the troll growls 'I want to eat you up', but the goat simply replies 'I want to eat *you* up', and knocks the troll into the river with the kind of uncompromising energy any three-year-old would admire. The three goats all live happily ever after.

It was that moment of reversal that I always loved the best: victim becomes victor, the goats are saved and the bridge is restored to its glorious function: connection rather than division.

Another Ladybird book on my shelf in those years was called simply *bridges*, with a small 'b'. On the front cover was a sketch of the iconic Forth Rail Bridge: three immense diamond webs of iron that stood between sea and sky as if connecting heaven and earth rather than Fife and Lothian. I knew the outline of the bridge well: it stood just a few miles from the house where I lived with my mum, dad and big brother. A special treat in those pre-school years was to take the train over that bridge into

Edinburgh: as we left North Queensferry I would watch as the bottom fell out of the earth and the train became sheathed in a forest of red steel that quite marvellously seemed to hold us up. Far below the tracks, tugboats like pond-skaters busied themselves over water so distant it appeared calm even in stormy weather.

The Ladybird book of bridges held within it a ten-thousand-year history of crossings. It began with a drawing of cavemen traversing a toppled log, moved on to precarious rope bridges and stepping stones, then in drawings of elegant simplicity explained the principles of arches, cantilevers and finally suspension bridges. The book was like a bridge to another world.

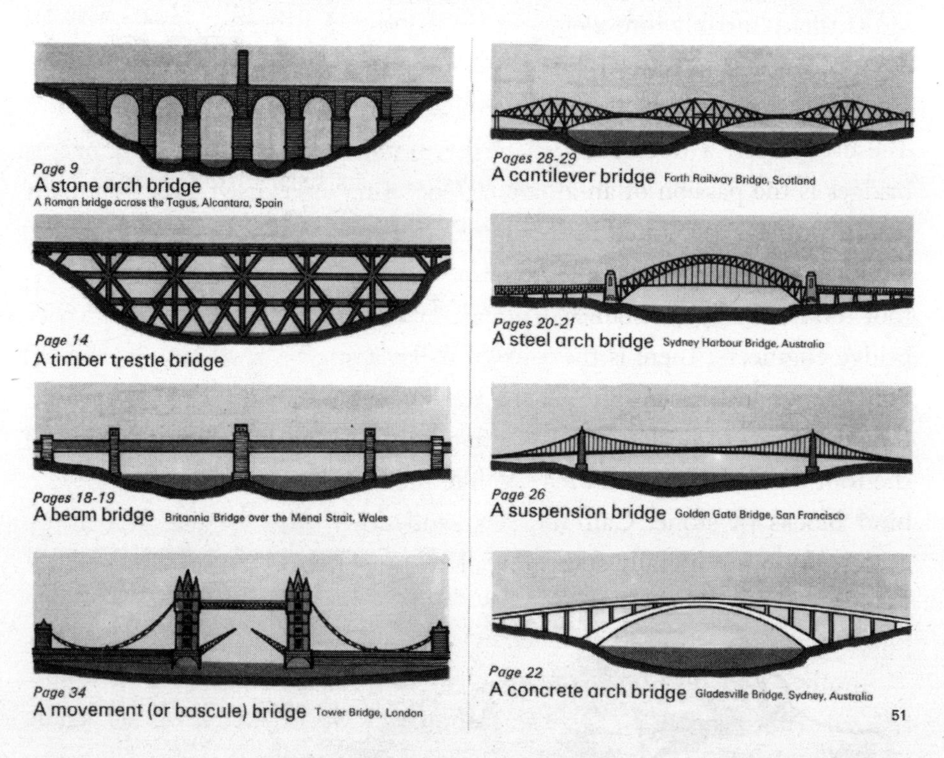

There at the most impressionable age it was made clear to me that bridges were among the finest of humanity's creations, with the power of connection but also of division. They were liminal spaces between departure and arrival; places not only of transition but of transformation. There were bridges with guard towers at each end, bridges that twisted on their supports to let river traffic past, bridges with decks that went up and down like trapdoors, and bridges with houses on them (how would it feel, I wondered, to sleep on a bridge, the sound of flowing water seeping into your dreams?).

Bridges had the capacity to change the whole meaning of a place, and could make the difference between life and death for the people who lived on either side of them. They were scales that could hold lives in the balance; they were objects of great power, but also of peril. The final page of the book showed an artist's re-creation of the Tay Bridge Disaster: shattered structural supports tumbled under the force of a gale, a train and its forlorn lights tilted horribly into a wind-whipped sea. To bridge some divides was to open oneself to reward, but also to expose oneself to risk.

These days I'm a doctor and a writer, not an engineer, and my love of bridges is the passion of an amateur. Doctoring lends itself to travel: many of the journeys recounted in this book were made in the course of medical work. When I taught anatomy I'd emphasise to my students that the human foot too is a bridge, sustained through four of the mechanisms adopted by bridge engineers. There is the arch of its structure, with a keystone-shaped bone under the ankle joint. There are ligaments like tie-beams that bind the abutments of heel and toe. There are tissues that buttress the arches of the foot across length and breadth, like the iron staples that masons use to bind blocks of stone. Calf muscles send down tendons like suspension cables. We quite literally stand or fall on our bridges.

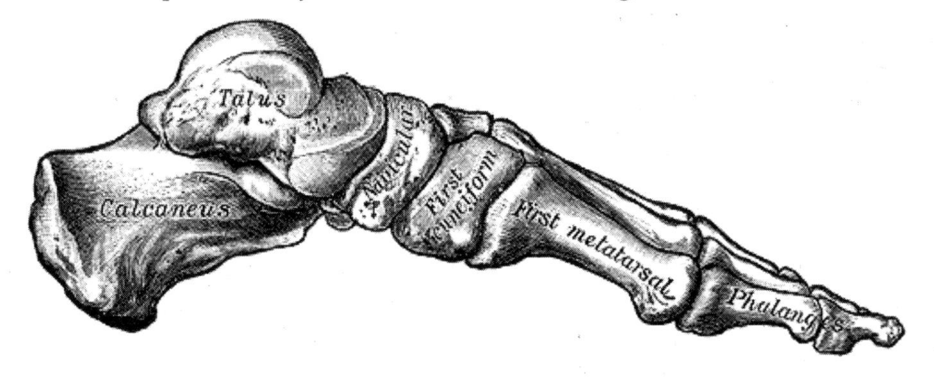

Our nose (of course!) has a bridge, and inside the brain there are many bridges: an arch of neurons that unites our left and right hemispheres, bringing together our creative and reasoning selves; the *pons* (Latin for 'bridge') where the upper, most evolved and quintessentially human parts of the brain connect to the lower parts that we share with more primitive animals.

My hope for this book is that it may itself be a kind of bridge from writer to reader, and from past into future. It's a memoir of my own restlessness, but also an exploration of the way travel can influence and inform perspective. It's a celebration of connection and cross-pollination. It has its origins in those little Ladybird volumes I read enthralled as a child, and in my many years travelling and admiring bridges, thinking about what connects people and what divides them. It isn't a history of bridges, but an exploration of their shifting meanings, actual and metaphorical, over two thousand years of engineering and almost fifty years of travel. Even the word 'metaphor' means 'a carrying across' – in the sense that the best metaphors build a bridge of meaning between speaker and listener. This book takes for granted the idea that bridges are among the most powerful of humanity's creations in terms of improving lives and circumstances, often emblematic of a culture, but also acknowledges their troubling martial history: some of the finest bridges in the world were intended to put one community at the mercy of another. It's arguable that bridge engineers have had a greater impact on the development of global history than military commanders.

Though bridges have been used for warfare as much as to foster civilian connections, it's impossible not to salute their builders. They take expertise gleaned by centuries of engineering and turn it to the most civil and civilising of purposes. They take the stuff of barriers and build from them crossings that speak not to our vulnerability to fear, but to our capacity to connect. They are our modern cathedrals in a sense, being the grandest of our constructions, a manifestation of what's best of our civilisation, embodiments of beauty and utility. Images of bridges are perennially popular because of the way they inspire imaginative crossings, dreams of escape and connection, loyalty and love. You can tell something about a society from the confidence and the extravagance of its great bridges, as well as from its more humble crossings of wood, rope and vine.

Most of those I've chosen to write about lie over or alongside borders, because the lines we draw around ourselves throw the power and possibility of bridges into relief. Though we are living through a golden age of civil engineering, and of digital interconnectedness, our own age seems uncertain about the value of strengthening and facilitating connections – for some, it's the very openness of cyberspace that is responsible for our era's surge in border anxiety. My own country has for several years now been focused on the hardening of its frontiers rather than the building of bridges. But for me, the latter are symbolic of our capacity to imagine a better future, to accept short-term losses in the hope of long-term gains, and as metaphors for the connections between peoples, bridges offer hope.

1980s

Chapter One

Bridges of Union

Tweed, UK
Union Chain Bridge: suspension (1820), 137m
Royal Border Bridge: stone 28-arch (1850), 659m
Border: Scotland–England

They do nothing for us at this end of the country. Nothing. This is like the back of beyond as far as London is concerned.

EILEEN BUCHANAN, BERWICK RESIDENT

On a childhood holiday driving from Scotland towards Yorkshire, the car pulled over into a lay-by. Mum and Dad leaned round to face my brother and me in the back. 'It's the border,' Dad said. 'Let's get out.'

He must have taken the picture. There's a sign, 'Welcome to England', the familiar blood-red cross on white. My brother is wearing a red holiday cagoule; I'm in blue, and our mum towers over both of us. Her hairstyle dates the image even more closely than the lettering on the sign, or the square Kodachrome print.

Was that the first time I crossed the Union Chain Bridge? I remember a single carriageway hung on suspension chains of iron, capable of bearing just one car at a time – though even with that burden of one, the deck would sway and bounce. Twinned boulder posts at each end ensure that only the narrowest of vehicles can cross, and the last time I visited, each post was scored deeply with the scrapes left by overconfident drivers. Flakes of paintwork had been left on the stone.

I don't remember standing for the photograph that first time I crossed the Tweed, but I do remember the river, the signs 'Welcome to Scotland', 'Welcome to England', and the strange sensation of crossing back and forward between nations with a single jump. The testimony of the bridge and the signs around it was that a line lay along the Tweed, an invisible line that was real nevertheless. It seemed a line of power, a magical line with the capacity to transform lives as well as landscape. A line of the law, a line of possibility. I marvelled that such a line could be conjured, and once conjured, so easily transgressed.

For almost a millennium the River Tweed was the frontier between Scotland and England. Over it at Berwick stands a trio of bridges, including one of 28 semicircular arches in stone. It was completed in 1850 as the final link in the Victorian train line between London and Edinburgh.

This new border bridge was built just a few years after Turner's masterpiece *Rain, Steam, and Speed*, which depicts a train hurtling towards the viewer on a railway viaduct high over the Thames valley. Turner's rail bridge is shadowy, ominous, though lit by reflections from the water below.

In the distance an old limestone bridge for horses and pedestrians shines as luminous as the clouds. A boatman bobs lazily, as if in counterpoint to the terrifying speed of the new age.

The Union Chain Bridge over the Tweed, completed in 1820, is the oldest suspension bridge in Britain still in daily use. Suspension bridges tend to be hung with cables today, but chains were more easily forged then, and it was designed by a navy captain renowned for his metalwork and his anchors. It's a bridge that seems spun from ship's rigging, more akin to the light and air that it stands in than to the quarries and foundries that made it possible. Oliver Riches, a bridge engineer I interviewed who has worked on building and maintaining immense bridges all over the world, spoke of the little Tweed crossing with enthusiasm. 'What a way to enter or exit Scotland,' he said. 'Now *there's* a lively bridge!'

In 1774 the Scottish engineer Robert Mylne made a sketch for the first-ever bridge of iron, commissioned by the Duke of Argyll at Inveraray, but that design was never built. Instead, the first iron bridge to be completed anywhere in the world was opened five years later, in 1779, across England's

Severn river. The Severn valley is rich in coal and iron ore, and would become the heartland of England's industrial revolution. The town that grew up around it is even today called 'Ironbridge'. When the Severn overflowed in 1795 the Iron Bridge was the only crossing to withstand the floodwaters. Other bridge engineers took notice: Thomas Telford, a peerless engineer of the period, began to work solely with this new material.

It's difficult to imagine now how revolutionary these metal bridges seemed when they were first constructed. To contemporary observers the use of iron seemed almost magical, certainly alchemical: the iron of the core of the earth heated in an inferno to create something entirely new – a structure that could float airily over a river. Other iron bridges quickly followed: in the Scottish Highlands, some of the most impressive were Telford's bridges over the River Spey at Craigellachie in 1814, and the Dornoch firth at Bonar in 1812. The poet Robert Southey, writing of his journey through Scotland, tells of a man whose father had drowned falling from a ferry near Bonar, and had ever since refused to cross the water by boat. He was nervous about the strength of Telford's new bridge: 'As I went along the road by the side of the water,' the man told Southey, 'I could see no bridge. At last I came in sight of something like a spider's web in the air. If this be it, thought I, it will never do! But, presently, I came upon it; and oh! it is the finest thing that ever was made by God or man!' With a little effort it's still possible to get a glimpse of the wonder the man experienced when confronted by a bridge of iron.

Perhaps in commemoration and reiteration of the frontier between Scotland and England, the stone-arched rail bridge over the Tweed is called the Royal Border Bridge. As if in protestation of the idea of borders, the bridge of steel rigging a little upstream became known as the Union Chain Bridge. Border and Union, connection and division, bridge and barrier: all these themes are combined in the naming of the crossings of the Tweed.

·

It seems a cruel absurdity to place our borders along rivers. Easy for legislators, perhaps, who draw up frontiers in oak-panelled rooms far from the rivers in question. But rivers give life to communities on both banks;

the water connects them more than it divides them. 'Berwickshire' is part of Scotland, but the town of Berwick itself is English thanks to the expedient, a few centuries ago now, of skewing the border away from the river, just downstream of the Union Chain Bridge, and five miles to its north. National claims on Berwick have swapped back and forth between the two countries somewhere between ten and twenty times in the last thousand years, but the town has been English for the last five hundred. An unofficial survey in 2010 found a majority of its townsfolk would rather it unite again with Scotland. Edinburgh is an hour's journey to the north, while London is more than six hours to the south. Scotland's more Scandinavian approach to public spending, with slightly higher taxation but higher welfare spending, was cited as one of the main reasons the locals wanted to lift the border like a piece of rope and drop it again a few miles closer to London.

Distrust of bridges and connection is deep in the British psyche: for the poet John Milton in *Paradise Lost*, it's the fault of bridges that humanity was expelled from Eden:

> a Bridge of wondrous length
> From Hell continu'd reaching th'utmost Orbe
> Of this frail World; by which the Spirits perverse
> With easie intercourse pass to and fro.

Sin and death came into the world by traversing a bridge from an underworld; *Paradise Lost* makes a good case for isolationism.

But on 3 June 1975, just a few weeks before I was born, Britain held a referendum on whether to stay within the European Community – a club of countries it had joined only a couple of years earlier. Margaret Thatcher, who would be prime minister by the time I was in nursery school, campaigned to stay in, wearing a jumper decorated with the nine flags of the Common Market. The referendum result was 67 per cent YES to stay in Europe, on a 64 per cent turnout.

Forty years or so later much had changed: those nine countries had become twenty-seven, the Common Market had become the European

Union, and the UK changed its mind, reasserting that traditional distrust. It chose to cut itself off, for the purposes of trade and migration, not only from the nearby French coast, but from the only country with which it has a land border: Ireland.

During the negotiations around the UK's withdrawal the prime minister of the time, Boris Johnson, announced plans for a super-bridge that would bind Scotland to Northern Ireland, as if in compensation or consolation for the more southerly connection that was being lost. But the strait that divides Scotland from Northern Ireland is deeper and more treacherous than that which divides England from France, and the plan came to nothing, as everyone (even the prime minister) knew it would.

The timing of the announcement was significant. Perhaps during periods of retrenchment behind borders, when literal and metaphorical drawbridges are everywhere being pulled up, people are comforted to think that the closure of bridges won't prove permanent. We want to hear that new connections will one day be laid across the boundaries we draw around ourselves.

Chapter Two

Bridge of Home

Forth, Scotland
Forth Road Bridge: suspension (1964), 2,512m
Borders: Fife–Lothian; Roman Empire–Caledonia

Scotland is not content to rely on past achievements, but is determined to remain among the leaders in all branches of technology. May this bridge bring prosperity and convenience to a great many people in the years ahead.

HM QUEEN ELIZABETH II

In my loft at home there's a newspaper clipping of a photograph of the Forth Road Bridge, with me and a thousand or so others setting off from its southern approach on a charity run. I was a couple of months shy of my thirteenth birthday; friends and family were to pledge a coin or two for every lap that could be made of it.

Runners stream past a caravan set up by the organisers, some frowning, some shouting, some grinning as they jostle together, each finding their space and their rhythm. By chance the press photographer has framed me dead centre: a Nike T-shirt, a sweatshirt tied around my waist, white sports socks pulled up tight, one of the smallest runners in the crowd. The camera has also caught the haze that day over the Forth estuary – the coast of Fife is almost lost in the mist of distance. As I ran twelve crossings of the bridge my feet felt winged, soaring over the river's cargo of ships and sailboats.

I was not bad as a long-distance runner, wiry and slight. Running suited me because of its solitude, and because of the way I could feel my heart beating in time with the rhythm of my feet, my lungs heaving in my chest even as my mind grew ever more airy and light.

From the hilltop behind my childhood home it was possible to see the topmost spires of the Forth Road Bridge, its towers painted the grey of doves' wings, and also of battleships. A Sunday afternoon's outing might be to take a walk over the bridge and back. If you close your ears to the traffic but open your eyes to the landscape, it's a walk to expand your mind as well as your vision.

A suspension bridge in the grand San Francisco style, it's an estuary-wide handfast of concrete and steel. Two immense towers more than a kilometre apart suspend twin garlands of cables that hang in graceful parabolas, like playmates swinging a couple of skipping ropes. Each cable is over half a metre thick, spun from 12,000 high-tensile wires and bearing 14,000 tonnes of weight – you could hang the Statue of Liberty off them, or three Eiffel Towers.

Two a half kilometres long, flanked by walkways and cycleways, with handrails like an ocean liner, it is fortified by gantries of trussed cross-beams to prevent it flexing in high winds. It was opened in 1964 by Queen Elizabeth as the Beatles' 'A Hard Day's Night' dropped to number 10 in the charts, and the Kinks' 'You Really Got Me' reached number 1. Its designers boasted that it would last 120 years, but the burden of traffic it has been obliged to bear has meant it needed reinforcement long before that. On that day in 1964 the Queen was driven north over the bridge, then sailed back again by ferry – the final passenger of a service that had been running for nine centuries. Three men died in the bridge's construction; several were saved by safety nets strung beneath decks. If you fell through those nets, it was a drop of 160 feet to the sea.

The bridge was astonishing in its day for the elegance of its slender towers, which had to be built in five sections. Its north tower was built first, 156 metres tall; it swayed even in light winds before a dampening system was laid into it. The anchors for each cable are buried 80 metres deep into the bedrock of both shores. Those hidden anchors confer the strength necessary for the openness and connectedness of the bridge – an apt metaphor for all those unseen footings that our connections rely upon.

The spinning of wire for its cables began in 1961 but was delayed by winter gales that year, which folded them in knots so tangled they had to be cut out. The decks of the bridge hang from its principal cables in 20- or 30-metre segments; metalled gaps between each segment allow for heat expansion in the summer. Cars driving over the bridge make a percussive rhythm as their wheels rumble over these metalled gaps, as if the bridge itself has a heartbeat and comes alive through the motion of those who cross it. Like the Golden Gate Bridge in San Francisco, and the Pont Neuf in Paris, the crown of the bridge is a renowned spot for suicides, but also for declarations of love: fixed there are grids of steel wire festooned with padlocks, each marked with hearts and the initials of lovers. It's a relatively new tradition, one thought to have started in Rome only within the last twenty years or so, underlining the way that the meanings we ascribe to bridges are always evolving.

About thirty thousand years ago the land on either side of the Forth estuary became entombed beneath the great northern European ice sheet. Fifteen thousand years later that ice retreated back towards the Arctic, unburdening the land, which began to ripple in geologically slow relief. Sometimes the earth would sink for a few thousand years, so that an inundation of the sea reached across almost as far as Glasgow (skeletons of stranded blue whales have been uncovered in the sediments west of Stirling). Sometimes it would rise for a few thousand years, to leave a series of raised beaches along the outer coasts. For long stretches of this time Britain wasn't an island, and much of the water we now have to bridge was still locked up in ice. Give or take a couple of river crossings, for those millennia you could have walked from London to Brussels with dry feet.

Almost two thousand years ago the Romans pushed the frontiers of their empire as far north as the Forth river, which they called 'Bodotria'. (Its equivalent frontier, at the other end of the empire, was the Euphrates.) This was their land under Ursa Major, the sign of the Great Bear – what the Greeks had called *arktikos* ('the bear'). It was with a strange sense of pride but also of alienation that I realised I had grown up in a place that for those Mediterranean explorers *was* the Arctic.

I have lived along the shores of the Forth for most of my life, but only once have I seen the aurora borealis flicker over its waters – a great arch of green light in the north, enveloped in darkness like an emerald ring tucked into the plush velvet of a jeweller's box. The Roman sentries who garrisoned this frontier of empire must have seen many such auroras, lights that for Scandinavians further north would become the bridge Bifrost, over which the gods could travel between heaven and earth. Later, medieval cartographers looked across the gulf of the Forth and called the region to its north Scocia Ultramarina – Scotland-beyond-the-Sea.

Not a great deal is known about the Forth in the centuries after those Roman colonisers left. In the eleventh century a Hungarian-born refugee princess called Margaret married a king of Scots and in 1071 established a ferry for pilgrims across the waters. That ferry was the forerunner of the three bridges that span the river today.

The settlements established on either side of Queen Margaret's ferry became 'North Queensferry' and 'South Queensferry'. An eight-hundred-year-old friary – the replacement for a much older wooden church – still stands by the old point of departure on the south side. It is even now in use, its paved floor sunk five feet below street level, its vaulted ceiling like an upturned clinkered hull. On the beach outside, children fish for crabs and tiddlers between rock sills that stood centuries' service as pilgrims' wharves. The medieval abbot of the priory was a kind of bridgemaster, determining who could cross and who would be denied.

As commerce, traffic and industry took the place of devotion, purpose-built piers were erected further east in the expanding town of Queensferry. In the late eighteenth century the Hawes Inn was constructed at the head of one of them. Sir Walter Scott wrote fondly of it in *The Antiquary* ('Well! we shall be pretty comfortable at the Hawes . . . it will be pleasanter sailing with the tide of ebb and the evening breeze'), and Robert Louis Stevenson too, in *Kidnapped* (There it stands, apart from the town, beside the pier, in a climate of its own, half inland, half marine – in front, the ferry bubbling with the tide.)

The oar- and wind-powered ferries Scott knew gave way to the coal and steam of Stevenson's day, which ceded in their turn to diesel. But the demand for those ferries relentlessly outstripped their capacity. The obvious solution – a tunnel – has been perennially proposed, and perennially ruled out: the rocks beneath the firth are lava-forged and too difficult to excavate, overlaid with many layers of glacial tills and silt. That said, there *is* a tunnel under the Forth further upstream, a service tunnel that once connected two collieries, wide enough only for a trolley of coal. But the river didn't suffer long to be crossed that way: after a few decades of being propped up, scooped out and pumped dry, in the late twentieth century the tunnel was abandoned to be drowned.

To cross any strait or watercourse, either by boat or by ford, has always been a risky activity, and was once much more so. Charon, ferryman of the Styx, demanded payment for access to the afterlife and to the bliss of forgetting; without that payment, the souls of the dead were trapped on the wrong side of the divide as ghosts, haunted by memories of their lives' misdeeds. In some ages and some cultures bridges have been seen as horizontal crossings to another world; in others, the crossing is vertical, with death coming as a fall from the precarious bridge of life. For Nietzsche, 'what is lovable in man is that he is an OVER-GOING and a DOWN-GOING'. As with death, birth too is a risky crossing between worlds, and the umbilical cord a life-giving bridge between mother and baby.

To confront the risks of river crossings, my own country developed a complex mythology of water spirits – 'kelpies' – that were to be appeased or avoided by travellers. Kelpie stories flourished until the military bridges of the eighteenth century made such myths unnecessary. Flowing waters have long been considered symbolic of the turbulence of life, and bridges, like ferries, emblematic of passage between its stages. As a child reading

my Ladybird books I marvelled at the power and possibilities of bridges, and even dreamed of living on one. As I get older I realise how much each of us lives *by* them.

Little remains to me now of that day running back and forth over the bridge at the age of twelve: a few flashing images of hot, exhilarating hours as I clocked up first ten, then twenty, then thirty kilometres back and forth between Lothian and Fife, Fife and Lothian. From the crown of the arch of the bridge the view was almost Olympian, and on that run I felt less earthbound than skybound. It seems to me now that I was running from the past into the future, from childhood into adolescence.

1990s

Chapter Three

Bridge of Possibility

Thames, England
London Bridge: stone 19-arch (1209), 267m
London Bridge: stone 5-arch (1831), 270m, relocated in 1971 to Lake Havasu, Arizona
Borders: Roman–Pict; south–north London

The old river in its broad reach rested unruffled at the decline of day, after ages of good service done to the race that peopled its banks, spread out in the tranquil dignity of a waterway leading to the uttermost ends of the earth.

JOSEPH CONRAD, *HEART OF DARKNESS*

In *Heart of Darkness* Joseph Conrad summoned an image of the Thames two millennia before the steamships of his own age, long before the tea clippers and the gunboats and the privateering galleons that powered the British Empire towards its domination of the world's seas. 'Sand-banks, marshes, forests, savages,' he wrote, 'precious little to eat fit for a civilized man, nothing but Thames water to drink.' A Roman trader, soldier or consul advancing up the Thames between swamps on one side and woodland on the other would feel as if savagery were closing in all around him, wilderness gnawing at his sense of himself. Before we have even reached page two of Conrad's story we're aware that this is going to be a very different tale of travel, and of imperialism, than those customarily told by his contemporaries at that high-water mark of Victorian empire.

Conrad imagined the Romans falling under a Britannic spell of darkness, the air and skies so utterly alien from the Mediterranean luminosity of home. His travellers would be disgusted by the Thames, but also enchanted: '[t]he fascination of the abomination – you know, imagine the growing regrets, the longing to escape, the powerless disgust, the surrender, the hate'.

And so Conrad's narrator begins his account of a faraway African river, into the darkness that lies at the heart of colonialism. And all the while, the 'traffic of the great city went on in the deepening night upon the sleepless river'.

Under yards of London mud and midden, and tonnes of brick and cobblestone, there was once a river flowing under King's Cross railway station – a ford over the River Fleet. There is even a story that Boudicca, queen of the local tribes who led resistance against the Roman invasion, died nearby after a battle in 61 CE. One legend has it that, like the Hogwarts Express, she rests somewhere between platforms 9 and 10. In the centuries following that battle a timber bridge was raised over the ford at the Fleet, and the area became known as 'Battle Bridge'. But the long-gone bridge didn't give rise to the 'cross' of King's Cross – that name came much later with a nineteenth-century statue of George IV, erected to preside over the fast-evolving hub of canals, railway tracks and a crossroads. London was to be the gathering point of an empire on which the sun would never set – a new Rome.

When I approached the station for the first time, at age 17, steel tracks seemed to tangle over one another, shining like fronds of kelp. Once out onto the platform, unseen voices announced the names of towns and cities I'd never heard of, in unfamiliar accents. Those accents sounded unexpectedly alien to me, *outlandish*, so unlike the accents of the mining villages of home, where emigration was more conspicuous than immigration. I had been invited to a 'science forum'; one of two hundred kids from all over the world sponsored by their home communities for a fortnight of freedom, attending lectures, going on day trips, seeing laboratories, partying with their peers, and having their eyes opened to the possibilities of a life in science.

On the street outside the station, plutocratic cars anointed the streets with expensive rubber. People everywhere, people *from* everywhere, people evidently from London but whose skin tones registered ancestral origins elsewhere, brought together by the branching, enriching, impoverishing networks of empire which all once funnelled to this place. People shouldering through the streets in a multiplicity of languages, fashions, attitudes. Euston Road was sclerotic with traffic; red and green homunculi flashed unheeded over each pedestrian crossing. The space above my head was dense with advertising hoardings, the air reverberated with car horns, the crowd seemed an organism with its own will, making its own demands.

Roads ramified from the central plaza like the spokes of an electric, illuminated cobweb of steel and concrete, and it seemed impossible for a moment that this city would ever end, and that it ever had a beginning. It seemed unlikely that metres beneath my feet the Fleet still flowed through estuarine clays, their origins in the melts of the Ice Age, and that once there had been marsh and field there, tangles of forest. That at one time a Roman trader or soldier or consul would have shuddered to think of the uncivilised darkness that reigned in this region north of the Thames, over such a forsaken part of the earth.

I arrived in London only to leave it again, descending into a somewhere-else – a white-tiled labyrinth of buskers and escalators, of musty winds gusting down starless platforms. An address on a scrap of paper took me to Camden Town. I emerged from the Tube station disorientated, the buildings too high, the canopy of clouds too thick to discern north and south from the position of the sun, at a plexus of five spokes of road. On a jostling pavement I set out northeast, startled by the city's plurality and possibility, its immensity of opportunity both to find and lose yourself.

Give or take a few fallow centuries, there has been a bridge over the Thames since the Romans first garrisoned Londinium, felling the timber from its dark forests to defeat the river, to cross over its waters and subdue the Britons. The Vikings hauled down one of many successors to that Roman timber bridge. In the twelfth century, inspired by knowledge of French bridges of stone like the magnificent bridge over the Rhône at Avignon (itself inspired by surviving Roman bridges), work began on one that would prove more difficult to demolish. Unlike the semicircular arches of Rome, each span was built in the form of an oval ('ogival'), making the cutting of its stones less uniform, but strengthening the crown of each arch. The spans were planned to be 24 feet in length (and resting on piers a mighty 20 feet wide) but had to be extended to suit the terrain of the riverbed. In the end its engineers were obliged to vary the spans from 15 to 34 feet in width.

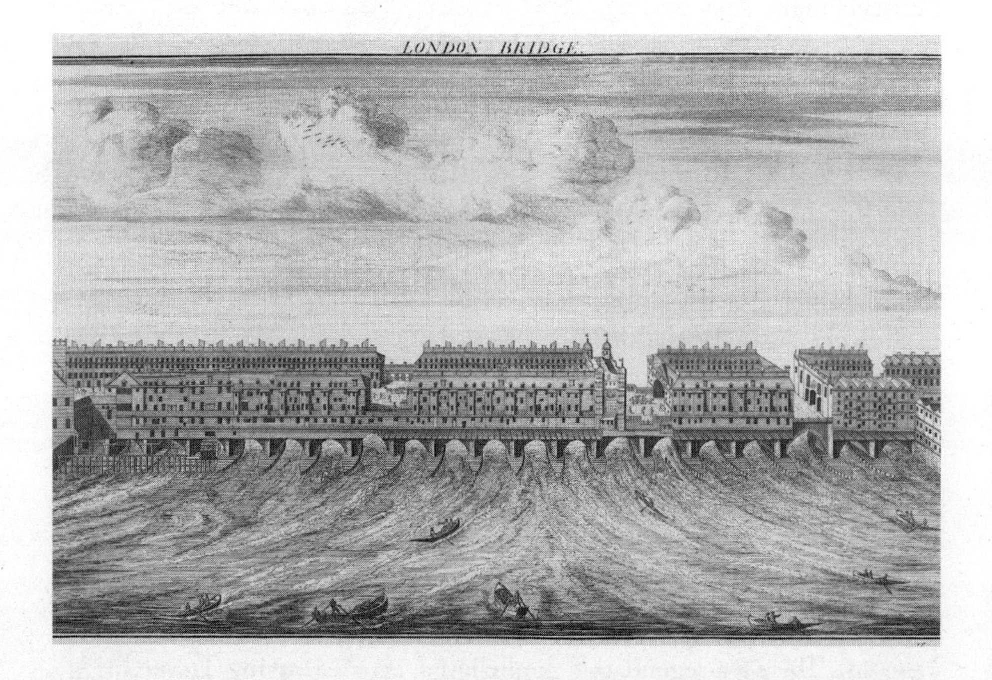

For the many centuries that London's medieval bridge stood in successive iterations it was top-heavy and hunchbacked, on staggered piers, with struts and props emerging from each side of it to support tottering storeys of housing. Its arches were so narrow in places that at certain moments of the tide the difference in water level between upstream and downstream could be as much as six feet.

We know from Shakespeare's correspondence that he lived near the north end of this bridge, and from contemporary engravings that every day on his approach he must have passed under a splayed arch of spears, a decapitated head impaled on each one – the bridge was not just a crossing, but a tool of social control.

Old London Bridge was dismantled in the early 1800s. *London Bridge is falling down, my fair lady.* It had been patched up and cobbled together so many times it was incapable of bearing the burden of a new age of commerce and interconnectivity. It had also become a barrier to shipping on the Thames: its piers were so wide, and the flow of water so turbulent between them, that passing under its arches was known as 'shooting the bridge' – sufficiently dangerous an undertaking that many passengers preferred to get out at a wharf upstream and walk around before rejoining their boat.

The successor to the medieval bridge, designed by the Scots engineer John Rennie, lasted from 1831 until the 1960s, when it too became unusable. The *Encyclopaedia Britannica* of 1910–11 said of it, 'as fine an example of a masonry arch structure as can be found . . . The semi-elliptical shape of the arches, the variation of span, the slight curvature of the roadway, and the simple yet bold architectural details combine to make it a singularly beautiful bridge.' A few years later T. S. Eliot in *The Waste Land* imagined it flowing with crowds of the undead.

In the 1960s Rennie's London Bridge was sold to the highest bidder: an American casino magnate who had it dismantled piece by piece and shipped to America, where it was rebuilt over a lake in Arizona to give his new town in the desert a cultural identity (and immediate global renown). There's a legend that he believed he was buying Tower Bridge, not the far plainer London Bridge. The journalists of London were bemused: 'London Bridge falls to the Apaches' was one headline.

At the north end of the medieval bridge was the church of St Magnus the Martyr, and at its southern end the cathedral of Southwark; each year in January the clergy of the two churches now lead a procession to the midpoint of the bridge's latest iteration. The replacement for Rennie's

London Bridge is a dull one of pre-stressed concrete, built in 1971, but that doesn't stop prayers being said for all who use it or whose work requires that they 'ply the waters'. I looked up pictures of past ceremonies online: bishops in gilt mitres and burgundy brocade led a crowd of worshippers and sightseers over the grey concrete. The blessing of the bridge seemed to offer a connection between the modern world and a more distant one in which rivers were honoured for the gods believed to inhabit them, a world before they became so polluted. Against a backdrop of café chains and office blocks, pebble-dash parapets and red London buses, the bridge was celebrated and consecrated, and the river appeased.

The science forum took place in 1992: there was no email, no mobile telephones, and payphones were usually out of order or clogged with queues. In lieu of phone calls, my parents had given me a sheaf of postcards, each one stamped and addressed for home. I had only to scrawl a few words on one every two or three days, and drop it in the post. My mum must have bought them cheaply, a job lot left over at the end of the tourist season: most of the postcards I sent from London were of the Forth Bridges. Recently she gave them back to me. The messages that were scrawled on them were telegraphic, cryptic, epigrammatic:

Met a few people . . . Tried and failed to find Leicester Square.
Went to Hyde Park with an Australian and a boomerang.
Been in Trafalgar Square at 1 a.m., wading in the fountain.

Alongside them I found some photographs. A park – Regent's, I think – fifteen teenagers from Ireland, Scotland, the USA, Malaysia, Australia, Canada clambering over one another to make a pyramidal bridge of bodies. Three tiers of youth with their floppy hairstyles, grinning for the camera, straining their limbs against collapse.

Those weeks in Camden changed my life, opening up a thousand new possibilities – of relationships between people and between ideas. Neuroscience and medicine became my subjects of study, and I developed a particular interest in the connections between brain and mind, and the many landscapes of mental health.

Recently a patient of mine told me about the work of a London mental health charity, Nafsiyat, based just up the road from the lodgings I was given in Camden. Its name in Arabic means 'psychology', 'soul' or 'mind'. The founder of Nafsiyat, Jafar Kareem, was a psychotherapist who in the early 1980s realised that the citizens of a city teeming with peoples, cultures and languages had need of a broader, more generous and accommodating psychotherapy than the traditional English or even European one. Mental well-being is inextricable from our social identity, and so to serve the needs of such a diverse community, a new approach had to be invented. Kareem called it 'intercultural therapy'.

This new kind of therapy would build bridges between therapist and client, taking for granted the notion that in matters of mental health it is necessary to acknowledge different worldviews, different languages, different values. It would seek to promote awareness of the many different layers and dimensions of clients' lives, and acknowledge how our conflicted, contradictory selves are inflected by background, upbringing and ways of seeing. After my patient told me about the place, I listened to interviews online with intercultural therapists, who see themselves as bridge-builders of the mind. For such therapists, it's not only a bridge between therapist and client that has to be constructed in the safe space of the consulting room, but bridges within the psyche of the client. Painful, disorientating, bewildering experiences can become islands of suffering. This therapy aims to reach that suffering with strong new bridges of understanding, which alleviate the pain we carry within us by allowing it an escape route.

Other disciplines of psychotherapy also use the metaphor of the therapist as a builder of bridges, and understand the mind as an archipelago of isolated elements in need of connection. Donald Winnicott, the paediatrician and psychoanalyst, believed that we need to develop a two-way traffic between the outer self that we present to the world and our innermost self; to leave those selves unbridged is perilous for mental health. From this point of view the goal of psychoanalysis is to broaden our human capacity for love and connection. The therapist Rosemary Gordon wrote in her book *Bridges: Metaphor for Psychic Processes* that 'bridging the different parts and tendencies within not only leads to greater integration of the personality but may enable patients to find more solid bridges to the people around them'.

London is a city of many divisions – social, economic, racial – and the Thames still represents a border of sorts: housing, transport infrastructure, job opportunities and even life expectancy are all richer, denser, higher in the north. For the Romans, the river was a frontier between worlds; for many of London's residents, it still is: the novelist Angela Carter, who lived in south London, described the Thames as 'the sword that divides me from pleasure and money'.

Bridges make the character of a city. In the mornings of that Camden fortnight I'd often stand on London Bridge to watch the water in motion with its cargo of water taxis, ferries and freight. Its neighbour, Waterloo Bridge, is notorious for sex and death, prostitution and suicide: there are many legends that swirl around it of fallen women redeemed by love as they contemplated their own end. And further upstream, Westminster Bridge: Wordsworth's excited poem 'Composed Upon Westminster Bridge' still held true, commemorating the transformation of perspective that bridges offer as their gift:

> Ships, towers, domes, theatres, and temples lie
> Open unto the fields, and to the sky;
> All bright and glittering in the smokeless air.

At the end of the science forum I fell back into my old life but changed, with a new consciousness of travel, of possibility, of the potential when people meet across borders. Bart's Hospital, the Institute of Electrical Engineers, the Natural History Museum – I had never been to places like these before. The awareness that they made knowledge grow, and that it might be possible to contribute a tiny part to the sum of human knowledge, was a tremendous gift of those weeks. London Bridge, Westminster Bridge, Waterloo Bridge – I had never been to places as relentlessly busy as these before. The crowds and traffic moving north and south, with the flow of the Thames beneath them, made me hunger for other cities, bridges, and other frontiers.

Chapter Four

Bridge of Vitality

Vltava, Czechia
Charles Bridge: stone 16-arch (1402), 515m
Border: Eastern–Western Europe

He made a human figure of clay, and left a small aperture in the lesser brain in which he laid a parchment with the unutterable name of God written on it. The clod immediately arose and was a man.

BERTHOLD AUERBACH, *SPINOZA: A NOVEL*

Somewhere in her books the Polish writer Olga Tokarczuk notes that in hostels and campsites and hotel lobbies the world over, travellers are forever scribbling in little notebooks. These journals are destined to pile up on shelves, on wardrobes, under beds, in attics. Who is going to read them? she asks. Isn't all this scribbling futile?

But the act of filtering reflections and impressions from the senses onto the page is surely valuable in and of itself. Who cares if no one ever reads them? Each one represents a kind of alchemy: experience transmuted in the crucible of the brain, then poured down through the channels of spinal cord, nerve, muscle, fingertips. The nib of a pen becomes a wand, conjuring a world that might otherwise have been lost to memory. It is in its own way a bridge.

Here's something conjured from one of my own travel journals: a castle in mist, a hillside dense with Bohemian forest. A trail meanders between trees in the full flourishing of summer. I glance to one side of the path as

I walk, and stop in surprise. There's a small clearing a few yards away, and in it a hole about three feet deep. Crouching inside the hole is a naked woman; smeared over her skin is soil from the forest floor. There are daubs on her face and handprints over her breasts. Her attention is focused on another young woman who is dressed in hunting camouflage, half squatting with her back to me, her long black hair hanging in a ponytail. She has an old large-format camera with a bulky lens; it rests on her thighs, and as she looks down into its viewfinder the naked woman begins slowly to rise from the earth, her hands unfurling towards the lens the way new buds reach skywards in spring. The photographer moves closer, the lens of the camera just inches from the woman's fingertips, trying to capture an image of dynamism and life as she rises from the clay.

The Vltava arises in the mountains to the south and west of Prague, along the rim of the Bohemian plateau. Bohemia is named for the Boii – a Celtic tribe who lived here two millennia ago, before the Slavs began to move west. For the Romans these were dark forested borderlands, an area of fluid frontiers. Julius Caesar had a wooden trestle bridge built over the Rhine to subdue Germanic tribes to the east and secure the eastern frontiers of Gaul. The construction took just ten days. It wasn't that he didn't have enough boats to transport troops; his bridge was effective propaganda, built to demonstrate the might and reach of the Roman army.

To the south, the borders of Czechia lie along the watershed between the Vltava, which flows towards the North Sea, and the Danube, which flows to the Black Sea. In the centuries since the passing of the Romans the Bohemian plateau has seen many struggles: between Franks, Slavs, Germans and Hungarians, and between Catholic, Orthodox, Protestant and communist. The river has, of course, flowed on regardless. A few miles north of Prague the Vltava empties into the Elbe, a river long considered the frontier between Eastern and Western Europe, which flows beneath Dresden, Wittenberg and Hamburg, carrying silts from the Bohemian highlands into the shallows west of Denmark – where they mingle with waters from the Thames.

Looking back from the distance of thirty years the river of time seems to quicken, undermining the bridges of my memory. Their mortar is crumbling and I'm left with only a few well-hewn stones to rely on.

It was the summer of 1993; the European Union existed on paper, but not yet in reality. On 1 January of that year Czechoslovakia had ceased to exist, divided in two; Czechs wouldn't become EU citizens for another 11 years. School was over for me, but university not yet begun. I worked 50 hours a week as a cinema usher, and for the first time in my life there was money in my pocket. I scooped popcorn, scooshed Cokes, ripped tickets, sold ice creams. I watched *Jurassic Park* 39 times, Schwarzenegger's *Last Action Hero* 27 times, Gary Oldman's *Dracula* 41 times. I shone my torch in the faces of troublemakers, swept up litter, and went in search down the sides of seats for dropped keys and wallets.

There were no cheap airlines. For £60 a bus company would take me and three pals from London's Victoria to the city of Prague and back again. It was a restless journey, fuelled by crisps and fizzy drinks; we passed Cologne, Frankfurt and Nuremberg by night and emerged into a new day deep in the Bavarian forest. I remember waking from a doze to hear an announcement that we were approaching the Czech frontier. 'Get your passports ready,' said the driver, stifling a laugh. 'Sometimes the Czechs check, sometimes the Czechs don't check.'

For almost five hundred years Charles Bridge in Prague was the only stone crossing of the Vltava. Like Old London Bridge it replaced an earlier timber crossing. The 'Charles' of its name is Charles IV, a boy raised at the French court who in 1346 – the year he was crowned king of the Romans by the Pope – joined the French in a war against England. This was a monarch at ease in the Slavic culture of the East but equally at home in the West, and Prague was the centre of his empire. In 1357 he laid the foundation stone of the bridge to stake Prague's claim as a city worthy of enjoying the kind of innovations then flourishing in the West.

It was a century of great European bridges, and Charles's bridge was part of the flowering of engineering that had seen the construction of the Ponte Vecchio of Florence a decade earlier – a much shorter crossing, but more daring for the shallowness of its arches (its architect, Taddeo Gaddi, had realised that with firm abutments on each bank it was possible to use a much shallower rise than had been traditional in Roman engineering). Charles lived long enough to lay the foundation stone of his bridge, but died shortly before its completion; his funeral procession was among the first burdens that its own shallow arches would bear.

PRAGUE. Old Bridge over the Moldau.

At its completion Charles's bridge was the longest of its kind anywhere in Europe. It is recorded of the bridge's builders that they had eggs baked into its mortar to imbue the stones with fertility and vitality. Alchemists and numerologists were consulted for spells and hexes to be incorporated into its stonework – the bridge was wreathed in magic. That was almost seven hundred years ago, and the bridge – through revisions and restorations – still stands. The seventeenth-century plague of baroquery saw its parapets festooned with statues, and by then its life-giving reputation had flourished and transformed. It was rumoured that on certain nights of the year its statues came to life and walked the streets, as protectors of Prague's newborn children.

There's a story I was told about Bohemia's place as the crossroads of Europe by Neal Ascherson, the writer, journalist, and expert on Eastern Europe during the Cold War. He spoke of attending a 1980s conference of Warsaw Pact countries in the Czech resort of Marienbad. The title of the conference was 'Building Bridges', and the Czechs were chosen as hosts because of their historical tradition of being a bridge between Eastern and Western Europe. 'The German delegate said "Bridges are good because they symbolise a higher Hegelian unity of self-realising subjects",' Ascherson told me. 'The Slovak delegate said "Possibly, but I don't want my country to be a bridge, because bridges are the first thing to get blown up in a war." The Russian delegate too was in favour of bridges and, looking around the room, he said "Bridges are very good, because it is possible to see who is crossing them."' What the Czech delegate said he didn't recall.

In my childhood atlas-wanderings, pre-1989, Prague had been somewhere beyond the possibility of reach – unapproachable beyond the Iron Curtain. But on those same atlas-wanderings it had always struck me how much of a Western city it appeared to be – a few miles from Nuremberg, to the west even of Vienna. The border between Western Europe and the communist East proved in the end insubstantial, rubbed out by the bureaucrats of Prague and of Brussels as easily as the frontier between the Czech Republic and Slovakia had been re-created, on 1 January 1993, at the stroke of a pen.

Of the city I have scattered memories: the glorious halls of its labyrinthine Metro; the frowning and soot-blackened saints along Charles Bridge, gesticulating at the heavens. The city seen from Petrin Hill, with its spires and bell towers like a page of exclamation marks. I remember pubs whose walls and archways betrayed their age – older even than the Habsburg Empire – and jackdaws chacking in the trees of the Old Jewish Cemetery; their wings, when startled, like applause. Before the Holocaust, Prague had one of the largest and oldest Jewish communities in Europe.

The word for 'jackdaw' in Czech is the onomatopoeic '*kavka*', or '*kafka*'. On one of my walks around the city I found the old building where Franz Kafka was schooled, training to become a crow in the great murmuration of Habsburg bureaucracy. A man with a Czech bird's name, who wrote in German a series of masterpieces he never saw published, stories like *Metamorphosis* and *The Trial* that have changed the consciousness of humanity. Kafka's Prague is a city without a name, and the heroes of his books seem to live without any knowledge of their past. Joseph K.'s city in *The Trial* is anonymous because its street names are mutable, lightly worn and ultimately irrelevant. Even Charles's bridge becomes simply 'a bridge'.

> All three of them now, in complete agreement, went over a bridge in the light of the moon . . . The moonlight glittered and quivered in the water, which divided itself around a small island covered in a densely piled mass of foliage and trees and bushes. Beneath them, now invisible, there were gravel paths with comfortable benches where K. had stretched himself out on many summer's days.

In one of Kafka's short stories, 'The Judgement', a young man is sentenced by his father to death by drowning, and promptly rushes to the bridge to carry out the command. Kafka emphasises that the river flows on without taking any heed of the trivialities of human life and death: 'He spied between the railings an approaching bus that would easily cover the sound of his fall, called out in a faint voice: "Dear parents, I have always loved you," and let himself drop. At this moment an almost endless line of traffic streamed over the bridge.'

It was the early 1990s; Bruce Chatwin's *Utz*, a novel about the life of a

Prague porcelain collector during the Cold War, had not long been shortlisted for the Booker Prize; Bohumil Hrabal's *Too Loud a Solitude* had just appeared in English – a novel about a man who pulps books banned by the communist regime. Milan Kundera's novels about the city were stacked on tables in every British bookshop. Kundera's *The Book of Laughter and Forgetting* includes an itinerary of the ways the conflicts of Europe's twentieth century had repeatedly reinvented just one of Prague's streets: Cernokostelecka Avenue (the Avenue of the Black Church). It was renamed after the First World War for Marshal Foch, then in the years of the communist ascendancy was again renamed for Stalin. Later still it became Vinohrady (Vineyards) Avenue. Like Kafka's, Kundera's novels keep returning to the river and its bridges, and so do his characters: the heroine of *The Unbearable Lightness of Being*, Tereza, 'wanted to see the Vltava. She wanted to stand on its banks and look long and hard into its waters, because the sight of the flow was soothing and healing. The river flowed from century to century, and human affairs play themselves out on its banks. Play themselves out to be forgotten the next day, while the river flows on.'

To walk in the city of the Prague is to feel observed by the stone eyes of sculpted saints, and by the hulking bulk of the communist-era caryatids that adorn its pillars and lintels. The entwined letters 'K&K' are still visible over many of its doorways, 'Kaiserlich und Königlich', remnants of Vienna's hold over the city which endured from 1526 into the twentieth century. The rock-built human city above stands in stark contrast to the gentle fluidity of the river below, with its soft banks of sinking silt.

There's a tradition that the silt of the Vltava was used by Prague's Jewish community for the manufacture of human effigies, or golems – beings of clay into which a simulacrum of vitality had been breathed, through an esoteric magical ritual inspired by the story of Genesis, Chapter 2: 'Then Adonai, God, formed a person from the dust of the ground and breathed into his nostrils the breath of life.' The golems would walk abroad at night, almost as the bridge's statues were believed to come alive.

Of the many stories of the golem that swirl around the old Jewish quarter of Prague that of the Rabbi Loew, of the later 1500s, is the most

notorious. The golem of Prague was a thing engineered, an enchanted hybrid between human, beast and clay. As a consequence it could be given only a limited, contingent kind of life, dependent on its keepers, but it could also escape the control of its master and run amok. To render the golem immobile it was enough to remove from its mouth a keyword stamped on a strip of metal – the creature would then slump inactive as if dead. In Loew's story the keyword becomes impossible to remove.

The story is as much about the perils of technology as it is about the nature of life and the danger of toying with it. As a cautionary tale it's in many ways similar to Mary Shelley's story of Frankenstein's monster. The Czech playwright Karel Čapek took this fear further in his play *Rossum's Universal Robots*, a kind of 1920s *Blade Runner*, which gave to the world the word 'robot' (derived from a Czech word for forced labour) and warned of a future in which artificial intelligence would threaten to wipe out mankind. Anxiety about humanity's frailty before the power of its technology seems perennial (and to be accelerating).

Later research has found no evidence in Prague of any sixteenth- or seventeenth-century golem tradition; it was likely a story invented in the early 1800s as the craze for the exploration of folklore traditions took hold across the Austrian Empire. The golem was pulled from the earth of cultural memory as a kind of robot warrior, given life in the realm of the imagination by embattled Jewish communities, at a time when anti-Semitism was again flourishing across central Europe.

Smetana, Dvořák and Janáček; Rilke, Einstein and Kepler: Prague's place as a cultural and intellectual centre of Europe is inarguable, though it wasn't necessarily so obvious for a boy growing up in Scotland before the fall of the Berlin Wall. In 1993 the city was reasserting its place as a crossroads of Europe, as part of that continent's heartland, and the encounters I had there, on either side of the bridge, built within me a sense of my own identity as a European.

In its bars I met a thin, angular Giacometti of a Pole who gave me a lesson in Czech: 'It's like Polish,' he said, 'except you have to laugh and sound drunk while you speak' – as good an endorsement for a language as I've heard. I met an Englishman with stubby broken fingernails who looked like a grave-robber; he told me he'd slept under bridges all across Europe before making the city his home – fluent in Czech, he now helped organise local folk festivals in Moravia. Of the encounters I recorded in my diary there was the French geneticist who worked on insect-resistant rice, the Swiss interpreter who spoke six languages with ease, the Finnish au pair who was working her way home to Helsinki capital by capital. I felt a sense of belonging among these Europeans, no matter the language barrier, though I'd have struggled to say exactly why. Beneath our conversations I sensed some fellowship, the hint of a shared endeavour, that I didn't feel with the Chinese tourist whose attempts at the Czech 'ř' sound were hilarious to everyone including himself, nor the trans-Atlantic trustafarian who insisted on buying all the drinks.

On my last night in Prague, on the way to my hostel, I stopped for a while halfway across Charles's bridge. I had it to myself; the river glittered with reflected moonlight, running north beneath my feet towards the distant North Sea, towards home. I knew that soon I would have to move out of the home I'd lived in for eighteen years, and start again in a city I barely knew.

But that night on the bridge I felt confident about the change ahead – my spheres of travel had been widening. A lively sense of opportunity hung in the air, of cross-fertilisation between cultures, of unity and fraternity, of connection and collaboration. Though the Czech Republic wouldn't join the Schengen area of passport-free travel for another decade, that agreement was already in place, and the city was buzzing with life and new ideas. And watching that river, I knew only that I wanted to travel on, past the headwaters of the Vltava, over to the valley of the Danube, into Asia and even on to the Americas. It seemed too much to hope that I might one day cross the Indus, the Yangtze, the Hudson.

Chapter Five

Bridge of Immortality

Tiber, Italy
Ponte Sant'Angelo: stone 5-arch (134 CE), 135m
Border: Italy–Vatican City

The first rule is, to keep an untroubled spirit; for all things must bow to Nature's law, and soon enough you must vanish into nothingness, like Hadrian.

MARCUS AURELIUS, *MEDITATIONS* 8.5

The train network of Italy must be the envy of, or an embarrassment to, many other states: an efficient ecology of regional, intercity and 'Red Arrow' trains that are punctual, economical, run on electricity, and so fast that there's little need for internal flights in that country, despite its generous length. Over the thirty years I've been visiting I've developed an intimate love for its many landscapes, the changes of its seasons, the music of its language, the reach of its history, and a deep affection for its people. And the food, of course; the whole world has fallen in love with Italian food.

When I first saw Rome at the age of 20 I was oblivious of all that. I remember the train from Florence sliding almost noiselessly over bridges and through tunnels, and depositing me beneath the great arches of Roma Termini station. You know what it is to arrive at a station like this (or its equivalents: Penn, King's Cross, Gare de Lyon): from standing in the aisle, bags in hand, impatient to leave, you are jostled to the platform. Neat brick walls; the air like a hot dishrag pressing down on the skin; light falling in shafts into the high, vaulted space, as if through cathedral windows.

Roma Immortalis, Rome, the Eternal City; at first glance it seemed a maze of crazily juxtaposed buildings and an impossible indigestion of architecture. Arched palaces beneath ugly concrete flyovers; Roman columns scattered over Etruscan stone slabs; everywhere pomp, grandiloquence and elegance. My provincial Nordic eyes were surprised by the atmosphere of authority and power, by the density of nuns on Vespa scooters, priests on motorbikes. Rome revealed itself as a deceiver: sometimes spacious and trustworthy, at other times hectic and treacherous. I was dazed by its contrasts: palaces crenellated in gold beside chapels gloomy with shadow, Renaissance magnificence alongside fascist braggadocio, cigarette stands built into the remains of classical temples. It was a surprise how thinly Catholicism was draped over the city, barely making an impression – even the Pantheon's conversion to a cathedral was unconvincing. Graffiti on every surface – *Eva ti amo! Sposami!* ('Eve, I love you! Marry me!')

At the border of Vatican City that graffiti stopped abruptly, perhaps from reverence, or from a protected cleaning budget. Inside the basilica of St Peter I paused at Michelangelo's Pietà, then hurried beneath successive ostentatious monuments to long-dead popes who did not appear to have been humble men – followers of Roman emperors, not of a carpenter's son. Each cold marble face looked down with disdain on the frail, fragile, forgettable flesh of the living tourists bustling at their feet. As a pilgrim from Europe's furthest limits I paid my respects to the spot where Charlemagne was crowned, and the throne of St Peter (a fake, built in the ninth century). It was possible to feel, for a moment, as if the power of Catholicism was undiminished by the centuries.

In a Roman youth hostel I picked up a dog-eared English paperback copy of the *Meditations* of Marcus Aurelius, the private diary of an emperor, allegedly written while on campaign in what is now Hungary and Slovakia, and thought by scholars to have been intended as advice for his son. In twelve short chapters it sets out a cosmic philosophy of personal responsibility within the churning and ever-renewing cycles of life; the stoicism it advocates is said to have influenced the development of Christianity, as well as the perspectives of prime ministers and presidents through the centuries. I still have that old paperback, and sometimes I reread it in the morning before beginning a day's work. It's a bridge not only to the reflections of a long-dead emperor, but to my 20-year-old self.

Walking the streets of Rome for the first time I could feel the power of the city's stonemasons, and how they'd made stone flow back and forth over the Tiber almost uninterrupted, three millennia of care in holding back the river in a set of graceful curves like parentheses. Every couple of hundred metres my feet would find a bridge, so that it was possible to move around the city almost heedless of the water that divides it east from west yet connects it to the high Tuscan mountains at the Tiber's source and the river's destination at the sea that brought it wealth – the Mediterranean. The river will prove more of an immortal than the work of any stonemason.

When the popes took on the mantle of the Roman emperors, one of many titles they adopted was *pontifex maximus*, the bridge between God and man – *pontifex* means 'bridgemaster' or 'keeper of bridges'. It's a title that recalls in its antiquity the origins of this city on the Tiber, at a time when river gods were believed all-powerful and to bridge them was an act of hubris open only to the most powerful of priests or rulers.

The word 'passage' comes from the Latin for 'steps', though it's long been used to describe journeys over water, and even for the stretches of water themselves. 'Passenger' comes from the same tributary of language. I was a passenger in this city, transported back and forth between its banks by its history and by its bridges.

If all roads were to lead to Rome, much of Europe had to be bridged. The first bridge over the Tiber was built around six centuries before the time of Caesar. Called Rome's 'first public work' the Pons Sublicius was of timber, and according to the ancient histories of Rome, in around 500 BCE it was destroyed by the defenders of the city in order to delay an Etruscan army. The commander of the city forces then was a man named Horatius Cocles: one story has it that he survived the wreckage of the bridge by swimming back to the Roman side; another that he was killed and fell into the water, sacrificing himself – an event commemorated by an archaic ritual called the Argei which persisted even into the reign of Augustus. According to Ovid, a procession of vestal virgins would pause at the bridge to throw in straw effigies of the founder-chieftains of the city, whose homeland had been Argos in Greece. To die in the Tiber was considered a substitute for the highest honour: to be returned by sea to the homeland of Greece, just as the dead would be ferried to Hades.

Another, darker tradition suggests that those straw figures thrown into the water commemorated the human sacrifice of Roman men over the age of 60 who, less able to fight, were of little use to a belligerent fledgling state on the make. The anthropologist and cultural historian James Frazer wrote of the tradition: 'So rooted in the Roman mind was the association of sexagenarians with a bridge and water death, that an appropriate and expressive word was coined to describe them – they were called Depontans, which means '"Down from the bridge with them!"'

The Ponte Sant'Angelo connects the Vatican with the Field of Mars and the Pantheon. It was built by the Emperor Hadrian's engineers in travertino and peperino stone, a broad reach across the Tiber to connect the heart of the city with the imperial mausoleum. It was normal for the early inhabitants of the Italian peninsula to site burial grounds on the far side of rivers, as if to divide the living from the dead. I wondered whether there was any overlap between the engineers of Hadrian's bridge in Rome and those of his famous wall in far-off Britannia – a wall I had by then

crossed hundreds of times on my way between Scotland and England. Hadrian's full name was Publius Aelius Hadrianus, and in Roman times the Ponte Sant'Angelo was named the 'Aelius bridge'. Later, after Pope Gregory had a vision of the Archangel Gabriel, the mausoleum was fortified into a castle – 'Castello Sant'Angelo' (the 'Castle of the Holy Angel')

– and the bridge took its name. For centuries after 1500 it was the city's execution site; its parapets, where tourists now lean out to take pictures of the river, were garlanded with the bodies of slain criminals.

I stood at the bridge's midpoint, balanced between one bank and the other, neither *urbis* nor *templum*. It has all the simplicity of the great Roman bridges, five spans each of eighteen metres in the form of perfect semicircles. A bridge engineer I met once explained to me that the Greeks were much better mathematicians than the Romans, but depressingly little survives of what they did with their knowledge in terms of engineering. It's perhaps from the Greek *póntos*, 'sea', that Latin derives its name for bridge, *pontum*, because for the Greeks, the sea *was* a connecting bridge, traversable by boat – a delicious paradox that sees islands cherished for their ease of connection, rather than the modern view that emphasises their isolation. The Romans on the other hand learned a few reliable rules of geometry as applied to stone, and repeated them across the Empire.

During the construction of a semicircular arch every stone can be pre-cut to precisely match the others – something that isn't the case with other kinds of arches, and which enormously simplifies the work of the masons. Bridge-building season was summer, to coincide with low water levels: an abutment would be created on each bank, and spans were lofted out over scaffolding, one arch at a time. When the waters rose in the autumn, building would pause until the following year. But with these simple rules (and their ever-replenishing supply of conquered slave labour), Roman engineers built the most enduring bridges on the planet, and transformed the landscape of Europe as well as its possibilities for trade, war and the extension of their empire. With lime and ash taken from eruptions of Vesuvius, they were the first to make widespread use of concrete. They also invented the cofferdam, an innovative technique of setting concrete underwater to found durable piers.

The architect Piranesi was fascinated by the Ponte Sant'Angelo: he made a famous drawing of its footings in cross-section that showed how much the relatively slender piers of the bridge, as seen emerging from the surface of the Tiber, constitute just a fraction of their true size. First of all the Roman engineers would have ordered two concentric rings of wooden pilings driven into the riverbed, in the outline of the cofferdam. Then they would have had clay packed in between the rows of stakes, as a means of making it watertight. Only then would all the water and mud from inside the cofferdam be bailed and shovelled out, and the riverbed excavated down to bedrock. If bedrock was too difficult to reach, close-packed pilings would be driven down so deeply into the oxygen-free silts that there was no danger of them rotting, and foundations of stone laid over these pilings. In the case of the Ponte Sant'Angelo, the piers descend to five metres below even the bed of the river, their foundations four times greater in area than they appear from above.

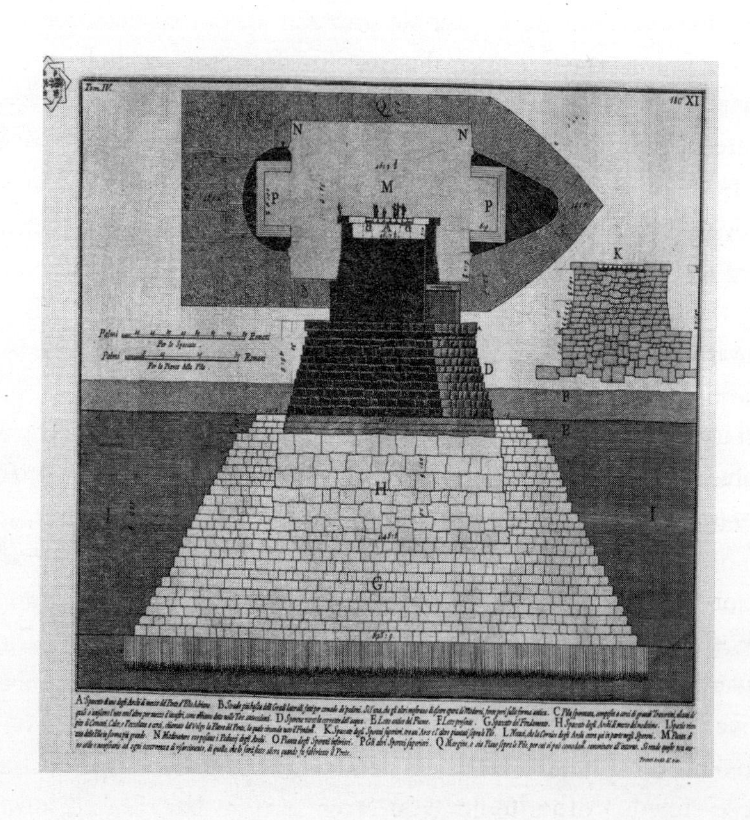

Each cofferdam and thus pier was shaped like a stone boat, prow directed upstream, cutting the flow of the water as if the bridge itself was in motion. Hadrian intended the bridge to his mausoleum to stand for eternity and speak eloquently down the centuries of Roman priorities: artistry, engineering prowess, military confidence. That the bridge of the deified emperor later connected the Vatican to the rest of the city bestowed upon it another meaning: pilgrimage. Iron balustrades were added in the thirteenth century; by the fifteenth, there are reports that a timber roof over the bridge was supported by marble columns. In the sixteenth, when it became an execution ground, the bridge was adorned with statues of St Peter and St Paul (the same impulse saw Charles's bridge in Prague adorned with saints). Medieval tradition saw bridges as alluring to both angels and demons; by the seventeenth century, the people of Rome protected theirs by arraying ten Bernini angels along its balustrades. A bridge first dedicated to the immortal memory of Hadrian had become a passageway to heaven.

Dante Alighieri transposed it from Rome into hell (perhaps the appropriate reaction of a loyal Tuscan): his *Divine Comedy* calls on the bridge's example to explain the then radically modern innovation of splitting traffic two ways. Dante had been in Rome in 1301, and heard how a year earlier, during the jubilee celebrations (in which greater absolution from sin than usual was on offer to supplicant pilgrims), one of his most detested popes, Boniface VIII, had insisted that the crowds going to and from St Peter's should use the bridge in carriageways. Dante transposed the crowd-control measures he'd encountered on the Ponte Sant'Angelo directly into Canto 18 of *Hell*, as Virgil leads him past the pimps and the flatterers:

> All of this was packed into the first ditch, along
> Whose bottom one group of sinners came
> Toward us while another naked throng
> Moved along with us, but with greater stride.
> Thus did the Romans establish order among
> Those crowds in the Jubilee year, when on one side
> Of the bridge half the people pressed

In the direction of the castle as they tried
To reach Saint Peter's, while all the rest
Went toward the mount. Horned demons lined
The bleak rock on both sides, and with cruel zest
Lashed their great whips on the sinners from behind.

That first time I saw Rome, and stood on Hadrian's two-thousand-year-old bridge between the city and the Vatican, between the cathedral of Western Christendom and the Pantheon of an older assembly of gods, I thought of Dante's horned devils stalking its parapets. What does it mean about our own age, I wondered, that we no longer build bridges to last two thousand years?

On the east bank of the Tiber black ministerial cars slid past beggars crouched and hooded; modern pilgrims in baseball caps and knee-length shorts jostled in the direction of the Vatican. It's surely futile as well as vain to expect an immortal legacy in stone, but Hadrian's comes close. The contrasts and contradictions of the city began to feel overwhelming, but also invigorating – so different from the Scotland I'd grown up in, the country the Romans had turned their backs on. Today Hadrian's most enduring legacies are his bridge, and his border wall.

Chapter Six

Bridge of Freedom

Venice Lagoon, Italy
Ponte della Libertà: stone 222-arch (1843), 3,850m
Border: Italian mainland–city of Venice

And let no man marvel that there are so many bridges, for you see the whole city stands as it were in the water and surrounded by water.

MARCO POLO, *THE MILLION*

August in the great bowl of the Po Valley, northern Italy, the heat of the sun trapped between the mirrors of the Alps and Apennines. There was smog over the cities dotted from west to east along the great river – Pavia, Parma, Padova – their names like an incantation. The earth of that valley is thick, loamy and fertile, good for growing crops and burying armies. Grapes swelled on vines, whispering to one another about how good they were going to taste. The only outlet for that great shimmering cinderblock of air is the valley mouth at the northernmost reach of the Adriatic. The air meets water there along a line between Venice and Ravenna.

Imagine yourself on an intercity train, itself as festooned with graffiti as each of the concrete bridges it slides beneath (Italian youth, glutted on the great beauty of their cities, despoil with their spray cans every *palazzo* and *ponte* they can reach). It threads its way between fields of maize and risotto rice and over the River Po, its waters so low that its bed appears more sandbank than water. The train's seats are old, slightly torn; some carriages are arranged in couchette bays of six. But they are clean, and the windows are all open to permit the grace of air in motion. Fellow passengers are dressed in stylish clothes and sunglasses no matter their budget. Signs around the windows: *'Non gettare alcun oggetto dalla finestra'*; *'Keine Gegenstände aus dem fenster werfen'*. You're 20 years old, with three years of university behind you, and your train is gliding towards Venice.

I was on my way to meet two friends from whom I'd parted a week earlier, in Florence. Vivek had wanted to pass the week in Rome, Justin had conceived a passion for Biarritz, while I had connected with a couple I'd met in Brindisi and together we'd determined to reach Amsterdam. In the absence of mobile phones and emails, the plan for our rendezvous was sketchy: 'Venice, at the Piazza San Marco, in one week.' With a rail ticket that offered limitless travel anywhere in Europe, I felt deliriously free.

Venice approached after a journey via Vienna, Zürich and Milan. In my rucksack was a Sony Walkman with one well-worn tape; from every third city I'd send a postcard home. I was five weeks into a journey on the rails, roads, sky- and sea-roads of Europe. The first two weeks had been in Poland travelling with a friend, Daniela: first with relatives of hers in Warsaw, then to her family's cottage by the lakes of Bryńsk, then walking in the Carpathians south of Kraków. Then on to Athens, to sleep rough with Vivek and Justin in the Peloponnese and work our way north. Each city we passed through felt like a spell cast on memory and the imagination.

The train met the coast and edged out over the waters of the lagoon; the dome of the sky was a great glass paperweight, pressing down. It seemed improbable there was going to be a city there to meet the track. Then church spires emerged over the eastern horizon, mirage-like. I looked south out of the window to see only water; the bridge itself felt as insubstantial as the city rising in the east.

The Ponte della Libertà, the Bridge of Freedom, was somehow raised out of the turmoil of Italy's mid-nineteenth-century wrangling between

Habsburg Austria, Napoleonic France and the nascent movement for Italian independence. Four kilometres of stone arches, resting on 75,000 wooden piles driven into the silts of the Venetian lagoon – a fusion of the old technology of cofferdams that would have been familiar to the Romans with the new technology of the railway. It was to be the terminus of 250 kilometres of rail track from Milan.

'In Venice,' wrote Jan Morris, 'the East begins.' Built at the fall of Rome's Western Empire, an island refuge from plundering barbarians, the city grew fat on its privileged position as a slave-trader between the Latin and Byzantine worlds. Its business model was to funnel captured people from Eastern Europe into the slave markets of the Mediterranean. The Slavic peoples had not yet been converted to Christianity and so under the laws of the day could be sold to Byzantine and Arab Muslim buyers without compunction. Venice made fortunes out of the discord between East and West, out of severing people from their freedom. From the air, its main islands seem to interlock like hands grasping in struggle.

Even today Venetians are adept at exploiting their unique position, though now they sell gondola rides, glassware and Gucci handbags to foreigners rather than selling foreigners themselves. Road and rail bridges may bind them to the mainland, but they are fiercely proud of their identity as Venetians, distinct from the Italian state. Venice is like a synapse, sparking with possibility between the Adriatic and the Po Valley, between agricultural and maritime worlds, between West and East.

Beneath the steps of the train station ran Europe's wateriest main street, bobbing with cigarettes and *gelato* wrappers. Crowds heaved along the walkways and bridges of the old city, pulsing through the Möbius labyrinth of it, towards and away from the heart of the city – San Marco's piazza.

From the square in front of the basilica I watched a ferry leaving for Greece. As its backwash reached the piazza's edge, gondolas were thrown up and down like piano keys. The cathedral, with its cupolas and domes, looks decisively of the East, more Orthodox than Catholic; on its façade it carries four equestrian statues looted from Constantinople at the time of the Crusades.

Vivek and Justin turned up right on time from their respective corners of Europe, and together we began to explore the city. Light glittered in arabesques on the underside of the many stone arches. Some of Venice's most famous bridges exist only in paintings, and as you walk the city it can feel like you're straying into the edges of that metropolis of the imagination. In 1747 Canaletto painted a Venice of architecture that had never been realised – among his many bridges was the unsuccessful design that Palladio had submitted for the Rialto.

Further on down the canal, at the entrance to the city's Arsenal, I sought out two immense stone lions looted from Piraeus, the port of ancient Athens. About a thousand years ago, Viking mercenaries from the fringes of the North Sea had carved runic graffiti into the lions' marble shoulders, back when it was Aegean light, not Adriatic, that glittered on them. Those Vikings were great travellers, reliable hired swords, but they left only the faintest of traces on these rich empires of the south.

One of my favourite accounts of Venice in English is that of Byron – the travel writer, not the lord – who arrived in the city in the summer of 1933. The fascists had recently completed a road bridge out across the lagoon, to flank the one of rail. In 1909 the Italian poet Filippo Tommaso Marinetti had demanded, in his *Manifesto of Futurism*, that humanity build 'bridges like giant gymnasts, stepping over sunny rivers sparkling like diabolical cutlery', and Mussolini had been happy to oblige.

Venice was to be Byron's starting point for a journey to Central Asia, written up in *The Road to Oxiana*. At the dockside he watched Jewish refugees from Germany embarking for Palestine, singing together of their hopes for Jerusalem and of their liberation from the anti-Semitism again flourishing in Europe. 'All towns are the same at dawn,' he wrote; 'as even Oxford Street can look beautiful in its emptiness, so Venice now seemed less insatiably picturesque. Give me Venice as Ruskin first saw it – without a railway.'

Much later, and on the other side of the Adriatic, I met Piero, a professional writer of travel guides and a true Venetian – someone who considers the front door of his home the one that opens onto his boat, not the one that opens to the street. 'In my grandmother's time, the population was 150,000; in my mother's it was 100,000, and in mine, now, it's just 50,000. Venice is dying,' he said, 'choking on tourists, who are killing the city they claim to love.'

The citizens of Venice have no say over the giant cruise ships and ferries that want to park beside the cathedral of San Marco; the power to grant their licences (and take the consequent port fees) is reserved to the Italian state. The archipelago of Venice has, Piero said, voted several times to separate itself from the continent, to again run its own affairs as a free city. But its voting boundaries include the larger population on the mainland at Mestre, and Mestre repeatedly says 'no'.

Venetians were campaigning for a law to insist that boats no bigger than those traditionally constructed in its Arsenal be allowed to approach the city. Tourists could continue to come, Piero said, but they'd be obliged to arrive in small numbers, in boats that wouldn't undermine the foundations of the city they've come to admire. For some Venetians the Bridge of Freedom has become the bridge of destruction.

I took a vaporetto east to the Lido, where there were short sections of beach open to the public without a fee, and paddled in water that looked out towards Istria, where the stone for many of Venice's great palaces and bridges was quarried: the Bridge of Fists (Ponte dei Pugni) and the Bridge of Breasts (Ponte delle Tette); the Rialto, the Paradiso and the much photographed Ponte dei Sospiri – the Bridge of Sighs. For criminals sentenced to death, to cross the latter was to leave one world and enter the next; the enclosed bridge meant that they were blinded from glimpsing the waters beneath them – water, the most life-giving of the elements.

Poised at the hinge of empires, at home in a city built on trade, Venetian merchants were honoured with places at the courts of Paris, Rome and

London, but also of Kyiv, Peking and Istanbul. Marco Polo's 24-year journey through Asia, in the latter part of the thirteenth century, is characteristic of the reach of Venetians of his age. To understand more closely the world Polo knew I went to see Venice's most famous map. Scholars describe it as the bridge between medieval and early-modern understandings of world geography.

The *mappamondo* of Brother Mauro, a Venetian monk, was created on the island of Murano around 1450, at the height of Venetian influence and so a couple of hundred years after Polo's time. It's more than two metres square, inked on vellum, and was the most accurate depiction of Europe, Asia and Africa then in existence. It paid tribute to Greek maps of antiquity, but infused them with traditions of Islamic cartography. It walks a diplomatic, Church-fearing line in explaining why it's no longer appropriate to place Jerusalem at the centre of the world. It knows the earth is a globe, and at its four corners there are inset maps of the north and south polar regions, the sun and planets, the elements of matter, and the Garden of Eden – which it daringly positions beyond the reach of earthly travellers, rather than on the banks of the Jordan or the Euphrates as was customary at the time. The accuracy of its Mediterranean coastlines is startling, but I was drawn in particular to the section describing the Forth estuary, my home.

Mauro wanted to eliminate the false belief that the Forth estuary extended across Scotland to the Clyde estuary in the west. Earlier medieval maps tended to divide Scotland in two, with a bridge at Stirling to connect its northern parts to its southern. Mauro made it clear instead that the border was partly the indent of those waters of the Forth and the Clyde, but recognised that there was also a belt of mountains that converged on an isthmus of land – the Ochil and Fintry hills. As a native of a city founded on money from the slave trade, it was the character of the Scots that most interested him: 'As it is shown, Scotia appears contiguous to Anglia, but in its southern part it is divided from it by water and mountains. The people are of easy morals and are fierce and cruel against their enemies, and they prefer death to servitude.' I smiled to think of the Venetian slavemasters encountering scowling, truculent Scots who, more than a century after William Wallace, would still rather die than lose their freedom.

At the age of 20, travel felt unintimidating, free, an expansion of mind as it kept pace with experience. It would be ten more years before I'd return to the Adriatic, journeying in the meantime to east and west Africa, to India, the Arctic, Antarctica, America north and south. But those six weeks on the rails of Europe opened a bridge in my mind onto a new landscape of possibility – a way of being that had at its heart curiosity about other places, other people, other ways of being and seeing.

Chapter Seven

Bridge of Division

Zambezi, Zambia
Victoria Falls Bridge: steel arch (1905), 160m
Borders: Zambia–Zimbabwe; European–African communities

The United Nations calls upon the United Kingdom of Great Britain and Northern Ireland to take all measures necessary to bring to an end the illegal racist minority regime in Southern Rhodesia and to effect the genuine decolonization of the Territory.

UN RESOLUTION 423

A bomb had exploded in the US embassy in Nairobi, and 213 people had been killed. Osama bin Laden was responsible, but it was a name that few people yet knew. Many Europeans were attempting to leave east Africa, and travel to the region was discouraged. I had been on placement as a medical student in a hospital in Kenya, and my train was stationary somewhere on the railway line between Nairobi and Lusaka. From Lusaka my aim was to take another train south to the Zambezi, across the famous bridge that hangs like an iron thread over a gorge of the Victoria Falls, the largest waterfall in the world. It's a bridge with one footing in Zambia and one in Zimbabwe. There was another student placement in the latter that I wanted to reach.

Outside the train window endless dun-coloured savannah, acacia trees, distant hills, a thin flanking track red with dust. Opposite me sat a young woman with strawberry-blonde hair and freckles; she held in her arms a souvenir wooden giraffe wrapped in oddly patterned paper that resolved at second glance into newsprint chequered with photographs of the bomb victims.

The Victoria Falls Bridge is in its own way a monument to failure. The failure of a diamond-mining tycoon called Cecil Rhodes to build a railway from Cape Town to Cairo, the failure of his dream that Britain would rule over all Africa (his ultimate intention: that it should rule the whole world), the failure of the white settlers of the region who dreamed of independence from Britain, the failure of those white communities to flourish in the country that carried Rhodes's name.

There were other failures – of the unrealised ambitions of independent African nations, of reconciliation between white and black communities, of understanding, of mercy, of imagination. But through all of these many failures, the bridge still stands, connecting Zambia to Zimbabwe – the countries once called 'Northern Rhodesia' and 'Southern Rhodesia'.

The Zambezi River flows for 250 miles across a plateau of hard basalt before reaching the Falls, where it tips over, thousands of tonnes a second, in a mile-wide sheet of water that is forced into a series of switchback ravines through softer rock. It's not a waterfall between two plains, but a waterfall into a mile-wide hole in the earth. By several measures it is the largest cataract in the world, twice the height and length of Niagara. It was known locally as 'the smoke that thunders', long before the Scottish missionary David Livingstone named it instead for his snub-nosed queen. Another local name was 'the place of rainbows'. It's as if the plateau is the edge of one world, and the river is an ocean ceaselessly falling through a crack in the fabric of the earth to become part of another. Funnelled into its tight canyon, the Zambezi flows on towards the Indian Ocean at Mozambique.

Livingstone certainly wasn't the first European to see the falls – it appears on early French and Portuguese maps of the area. He came upon it in 1855, and only five decades later the bridge across the gorge was complete. The railway is flanked by roads and walkways, standing four

SOUTHERN RHODESIA. AERIAL PHOTOGRAPH OF VICTORIA FALLS.

hundred feet above the river's surface, so close to the falling water that the engineers, of the firm Freeman Fox & Partners, had to take into account the cooling effect of its spray on the metal. The bridge took just over a year to build, constructed from parts brought in by the trains it was destined to carry – a 160-metre arch of trussed steel, bolted into concrete foundations.

Cecil Rhodes didn't live to see it – he died at the age of 48, three years before the bridge was complete. The joining of the bridge was reported in the local newspaper, the *Bulawayo Chronicle*:

> The junction this week of the two arms of the great steel arched bridge . . . is evidence not only of British colonising enterprise, but of the skill and pluck of the British engineers, alike in design and construction. We have thus seen what was a generation ago an unexplored region subjected to the commercial and civilising influences of the railway engineer; and as the gorge spanned by the bridge was

> one – perhaps the greatest – obstacle to that great scheme of Cecil Rhodes for opening up Africa by a railway from the Cape to Cairo, the close of the steel-work is an event of far reaching importance.

A reach of fewer than 200 metres was of far-reaching importance. But the bridge was seen as the most challenging rivet in a continental span that would funnel on steel wheels the wealth of the continent down to its ports and into the ships of a seaborne empire. A vision that spanned from Suez to Soweto but passed unseeing over the peoples of Zambia and Zimbabwe.

Seventy years after the bridge's construction, its liminal, in-between nature led to it being chosen as the site of a peace summit – held in a train carriage poised halfway over the gorge. On one side were delegates from the Rhodesian government, white men who had declared independence from Britain only a decade before, and who were fighting a civil war to maintain Rhodesia as a white-governed state. On the other side were delegates of the African National Council under Bishop Abel Muzorewa, who sought democratic representation of all the nation's peoples. Nine and a half hours of talks proved fruitless. The leader of Southern Rhodesia, Ian Smith, refused to grant diplomatic immunity to African nationalist attendees. Bishop Muzorewa accused him of inflexibility and a lack of goodwill. The summit ended in failure; the civil war intensified, and less than five years later, Robert Mugabe was elected leader of Zimbabwe.

In the dining car I sat with two students heading home from university in Nairobi, both marijuana-stoned into incomprehension; they sniggered at my attempts to make conversation. The train was delayed by rocks on the line. Outside in the dusk little children ran back and forth beside the track, calling out, empty palms aloft. The train began to shunt forward, past a boy cross-legged at a hut reading schoolbooks; a herd of impalas; another of zebras; pink and sapphire birds that flashed vivid against the darkening grassland. Earlier in the day, women in traditional beads and headdresses had jogged alongside the slow-moving train, holding up drinks to sell at the windows. A French couple in the next carriage caused a scrum throwing balloons to the children.

The following morning, the train still had not moved. A man with teeth like the tines of cogs, blunt and interlocking, sat opposite me at breakfast. 'This continent is no longer yours,' he said to me, pieces of egg falling from his mouth as he chewed.

'I never thought it was,' I said.

'Your people thought it was theirs.'

I chewed on in silence, wondering how I could ease the animosity between us. 'I'm not here to tell anyone what to do,' I said. 'I'm here to learn.'

We were joined by a photographer from Harare, who until the downturn in tourism had made much of his money by providing images for glossy brochures. He too singled me out as a representative of all Europe. 'Why are you people not coming?' he asked me. 'Are you all *scared*?' His daughter fixed me with a resentful stare. I shrugged, and muttered something about the bomb in Nairobi.

Some of the people I spoke with didn't bother to disguise their hatred of me, or of what I represented to them; others were simply curious. Why had I come to their country? Descendants of white settler families let slip a similar level of resentment, but its origin was different: not that I or my ancestors were the beneficiaries of colonialism, but because I wasn't more of an ally to their cause. Confused and embarrassed, I found it easier to stay silent.

Long train journeys lend themselves to big books, even more so when those train journeys are interrupted by rocks on the line, maintenance issues, bribes to be paid, cancelled connections. Before I left for Kenya, Vivek had given me Saul Bellow's *The Adventures of Augie March* to read on just such a journey. Gazelles cantered alongside the train and once, in the distance, a lolling group of water buffalo; all the while I read how Augie March ricocheted around Depression-era Chicago, then Mexico, trying on different styles of life like so many costumes, only to discard them when their demands or complexities began to rankle. At the age of 22 I recognised the way Augie March obsesses over vocation and destiny, and wondered too how it might feel to be driven by a sense of right and wrong, by a social purpose or by ambition. 'I always believed that for what I wanted there wasn't much hope if you had to be a specialist, like a doctor or other expert,' Bellow has Augie say; 'and besides specialization means difficulty, or what's there to be a specialist about?' He adopts the antithesis of this difficulty as a mantra for life, 'Easy or not at all', and resolves that for every crossroads in his life he'll choose the easier road.

The news from Zimbabwe wasn't promising: there had been food riots, and martial law had been declared. Non-essential travel to the region was discouraged. As I struggled south across a continent so vast that my own country of Scotland would fit into it a thousand times, Augie March's words began to ring in my mind at the most unexpected moments: 'Easy or not at all.'

A few hours later and the train was still stationary. A trickle of passengers began exiting, dragging their plush luggage across the savannah towards a village just visible on the horizon. I joined them: shouldering my rucksack past huts with earthen walls, barefoot children, dark doorways and wandering chickens. At the highway no freight trucks would stop, but a nine-berth taxi juddered to a halt. I crammed into the back where two women, Lucy and Monica, shifted themselves to reveal a narrow sliver of seat into which I could insinuate my hips. Prices were agreed; the taxi set off, only to drive to the village marketplace, where the driver got out without explanation. I was glad of the company of Lucy and Monica, which began to feel like protection; they took immediate charge of our new, intimate and temporary community by insisting that everyone refuse to pay the driver until he accepted the official rate, and that we also refuse to get out of his taxi. It worked: the driver returned, and began to drive us to the nearest town.

Colossal trucks lurched through lakes of slurried mud. The taxi pitched and swayed through holes in the road as if we were at sea, the chassis grinding against rocks, and with sadness I realised that I didn't have the necessary energy, or the will, to reach Victoria Falls. The love of travel I'd developed in Europe had found its unhappy limits.

In the nearest large town that night, as dusk fell over the plains of east Africa, the Falls still lay hundreds of kilometres over the southern horizon. 'Easy or not at all,' I said to myself, as I dragged my backpack into the foyer of a hostel and began to think about finding a ticket to the north. 'Easy or not at all.'

Chapter Eight

Bridge of Poetry

East River, USA
Brooklyn Bridge: suspension / cable-stay hybrid (1883), 1,833m
Border: Manhattan–Brooklyn

Cross from shore to shore, countless crowds of passengers!
Stand up, tall masts of Mannahatta! stand up, beautiful hills of Brooklyn!
WALT WHITMAN, 'CROSSING BROOKLYN FERRY'

From twenty thousand feet Long Island looked like a smear of embers, a firepit waiting for a fakir's walk. At school I'd had to study *The Great Gatsby*, and a line of that book came back to me, of how Gatsby's grounds at West Egg must have seemed at the moment of its colonisation by Europeans. Confronted with the American continent, they came 'face to face for the last time in history with something commensurate to [their] capacity for wonder'.

The border guards at JFK had mercifully ignored me and after the customary hour or two's wait, I went down into the maze of the subway, into Manhattan's stink of opulence and dereliction. The map of its connections looked like a baffling circuit diagram, and I wondered how I was ever going to find my way around. But New York is a city so instantly recognisable from American TV and movies that it has a familiar wallpaper all of its own – though that unexpected familiarity was itself unsettling. It was my first time in the United States of America, and the streets were as electric as I'd expected – with the scream of ambulances, anger and anxiety, distress and distraction. The skyscrapers were like monstrous, magnificent cliffs, sprawling and unsettling, the streets between them turbulent and chaotic but also seductive.

I was to stay in a hostel on East 27th Street, not far from Fifth Avenue. Every kind of body habitus was on parade in the street outside, from startling ugliness to enchanting beauty; walking a few blocks was a lesson in diversity and unity, each face in the crowd like a card from a tarot pack. I was 23 years old, and the surging masses of people, the relentless flow of life and minds, of intention and desire, appealed to a part of me I wasn't aware of before.

I was a medical student then, trying to persuade Bellevue Hospital to accept me on its exchange programme. To get there each morning I crossed parks filled with broken, filthy men bellowing to the sky. Other homeless men lurched threateningly around the sidewalks as if demanding to be noticed, but the crowds simply opened and closed around them, the gaze of each pedestrian averted. Some grinned as they rattled their cups for change; a middle-aged woman stuck her tongue out at me and blew a raspberry. On Fifth Avenue a young man crippled by some disability staggered along the pavement, buckled and bent over, screaming in a voice that was desperate and insane. It was a city of surpassing excellence and of pervasive wreckage; avid for money, beeping and flashing like a fruit machine.

I spent days pacing the sidewalks of Manhattan and Brooklyn. Among those rivers of people flowing by night and by day, I remember a tiny, desiccated elderly white woman, supported at each elbow by a black maid; a beautiful young Asian man walking Sixth Avenue singing operatically. Roadblocks of police at the United Nations, and around that block, an agitation of protesters. I fell in with a crowd of Irishmen, and went out on a night of booze and blues in Greenwich Village. But on all those walks I kept returning to the bridge.

The poet Hart Crane declared that Brooklyn Bridge was the most beautiful bridge in the world; he loved its 'frozen trackless smile':

O harp and altar, of the fury fused,
. . . thy swift
Unfractioned idiom, immaculate sigh of stars

He saw America as a kind of bridge between Europe and the Asia Columbus thought he had reached, and his poem is a great experiment in the power of metaphor in the sense of 'carrying across'. Crane lived in the Brooklyn Heights apartment where the bridge's engineer, Washington Roebling, once lived – an apartment that the engineer had chosen for its panorama over the steadily ascending Babels of the bridge towers. Roebling had taken over the project following the death of his father John, who had died from tetanus sustained in a crush injury to the foot – squeezed between a boat and a wharf. To be killed by one of the boats that his bridge would supersede was an irony that came to dominate the legends of Brooklyn Bridge. The crippling of Washington Roebling was another.

To sink its towers a new type of caisson had to be invented. Colossal hulls of timber were constructed then tipped upside-down onto the riverbed. Air was pumped into them under pressure, to keep water from seeping into the upturned bells while the workers hewed away at the East River's bedrock – a hellish mixture of clay and granite. As the caissons dropped the pressure inside them grew, so that workers emerging from the pier foundations began to succumb to 'the bends' – or decompression sickness. It's a condition now known to be caused by the release of bubbles of nitrogen within the capillaries of bone and nerve – nitrogen that dissolved in the blood under pressure, being released once the workmen were taken back up to the surface. But it was then unknown why men would surface from these pressurised spaces in agony, paralysed or even dead.

Washington Roebling was to experience this decompression sickness first-hand on two occasions: the first time he recovered, but the second time he sustained a spinal cord injury that paralysed his legs. His wife Emily took over the day-to-day work of supervising the construction of the bridge, and in doing so was revealed to have been pivotal all along – one of the few instances in the history of bridge engineering in which a woman has been given a degree of credit for her work. As part of the bridge's triumphant opening ceremony (14 tonnes of fireworks, 13,000 invited guests) Emily Roebling was one of the first in a procession to cross the bridge, and carried with her a rooster – emblematic of victory.

For Hart Crane, the bridge was a symbol of American audacity and ingenuity, a beautiful ideal, a manifest emblem of transcendence. It was completed in 1883, its new design combining the use of innovative new suspension cables rather than chains or pieces of cast iron. The vertical hangers of the bridge fall from those cables to the deck, which is buttressed by cable stays that run directly from the towers. It's those crossed suspension hangers and cable stays that give Brooklyn Bridge its strange resemblance to a harp, belt-and-braces engineering that is taking no chances. It is unusual, too, in having its walkways positioned centrally and elevated above the vehicles, so that as you walk across, you're soaring over both river and traffic. The footsteps of pedestrians beat a constant tattoo on the taut drumskin of its deck.

In those first weeks in New York I walked back and forth over the bridge many times, trying to commit a stanza of Marianne Moore's poem 'Granite and Steel' to memory – a stanza which invokes the tenacity of John Roebling in transforming a design as elliptical and romantic as a rainbow into a solid monument of steel and stone. Beneath me the East River flowed on as smoothly and elegantly as the drapes on the Statue of Liberty, past the tip of Manhattan and the twinned towers of the World Trade Center.

The whisper of city traffic merged with wind in the cables, a drawn-out exhalation. The bridge was so innovative, its cables so insubstantial, that when it was first used people feared it would break – within a short time of its opening there was a stampede to get off it, triggered by a rumour that it was beginning to crack. The circus impresario P. T. Barnum broke the spell – by leading a cavalcade of twenty-one elephants, seven camels and ten dromedaries across the bridge from Brooklyn to Manhattan.

New York is an Atlantic archipelago; like Italo Calvino's city of bridges it lives and breathes by them. Only the Bronx is part of the continental

United States. To visit the city is to be stunned again and again by the scale and splendour of the bridges that recur in every vista, every song, every piece of art – from the kick and snap of Simon and Garfunkel's '59th Street Bridge Song' to F. Scott Fitzgerald's New York as seen from the Queensboro Bridge: 'always the city seen for the first time in its first wild promise of all the mystery and beauty in the world'. The Hell Gate Bridge of 1916, a prototype of Sydney's Harbour Bridge and just as impressive, is in New York swamped by competitors. Even bridges that elsewhere in the world would be tourist attractions are here overlooked. Just as the sunny side of these bridges is an invitation to transcendence, they have their dark underworld too – a medieval cosmology played out in civil engineering. In Italian to be *sotti i ponti* ('under the bridges') is to have fallen into a dark underworld of crime, homelessness and degradation. Beneath symphonies of decks and cables lies a liminal world of approach roads and flyovers, among poorer neighbourhoods and fenced-off lots. Land-hungry New York can't develop much of the real estate under these bridges, so they become the habitat for a twilight citizenry – New Yorkers who have nowhere else to go.

I love to arrive in New York from the west, through the doubled tiers of the George Washington Bridge – built in 1931, its towers like the teeth of a sky-cutting saw. I remember a bus driver switching on the intercom as we crossed the bridge: 'Here it is, folks, New York. I hope you've all got a looooot of money, 'cos you're gonna need a looooot of money.'

More than a hundred million vehicles pass over the George Washington every year, urgent, impatient, each one moving as if it's being charged by the minute for the use of the roads. The architect Le Corbusier thought it the most beautiful bridge in the world, 'the only seat of grace in the disordered city'. 'When your car moves up the ramp the two towers rise so high that it brings you happiness; their structure is so pure, so resolute, so regular that here, finally, steel architecture seems to laugh.'

* * *

George Washington Bridge and Hudson River, New York
15

On one of my walks to Brooklyn Bridge I came across a torn chain-link fence, and behind it an encampment settling in for the night under tarpaulins. So much for the American Dream. In his poem 'Crossing Brooklyn Ferry', Walt Whitman wrote of this same stretch of the East River that it seemed to concentrate all the wonder and possibility of the city:

> And you that shall cross from shore to shore years hence are
>   more to me, and more in my meditations, than you might
>   suppose . . .
> The glories strung like beads on my smallest sights and hearings,
>   on the walk in the street and the passage over the river.

For Whitman, who moved from Long Island to Brooklyn as a child and worked often in Manhattan, the Brooklyn ferry's motion back and forth was symbolic of American vigour, dynamism and transcendence; for Hart

Crane, the bridge standing over the same passage of water represented the same.*

Whitman called upon the reader to imagine the East River far into his future – our present – and gasp with him at the changes to the city he'd never see but had faith would come to pass. His poem is layered with American myth-making, taking the ancient idea of ferries to the underworld, passages of transcendence, and bending them to fit his new American aesthetic. In his day the wharves of Manhattan bristled from the southern end of the island; there were ships berthed along the lower Hudson that had sailed from all over the world. Herman Melville wrote of it: 'There now is your insular city of the Manhattoes, belted round by wharves as Indian isles by coral reefs – commerce surrounds it with her surf.'

The population of New York was in Melville's time swelling fast – it grew from 500,000 in 1850 to 3,500,000 in 1900. In 1904 Henry James, the novelist and Bostonian, visited the city researching a book that would be called *The American Scene*. He was shocked by its flood of immigrants (an unprecedented 1,004,756 arrived legitimately in 1907 alone, the year *The American Scene* was published). Yet James was knowledgeable enough about his country's history to reflect that even in this long-inhabited city, called New Amsterdam by Dutch settlers, then New York by Englishmen, the land was held in '*un*settled possession'. A country founded on immigration and the expropriation of land couldn't close itself to immigrants, build walls around its borders and say 'enough', he wrote – New Yorkers had no choice but to accommodate waves of new arrivals. Almost everyone in America was an alien, he added; New Yorkers didn't own America, and so they could never be dispossessed of it.

Medical school was almost over for me; soon I'd need to apply for internships and decide the direction of my future career. Despite my pleading

* Crane would die at the age of 32, leaping from the stern rail of a steamer in the Gulf of Mexico. 'Bequeath us to no earthly shore,' he'd written, 'until / Is answered in the vortex of our grave / The seal's wide spindrift gaze toward paradise.'

Bellevue wouldn't offer me a student placement, and I made plans instead to travel to Bethesda, Washington DC, to spend time with a group of neuroscientists at the National Institute of Mental Health. Following that placement I'd return to Scotland, to a rural hospital in the Highlands. I was reluctant to leave New York – its energy was intoxicating – and on the night before leaving the city I walked to the edge of South Street to look once more at Brooklyn Bridge.

Roebling's Gothic towers, *harp and altar, of the fury fused*, and those cables – *Climactic ornament, double rainbow* – could be seen as those of a new kind of church, one that worshipped a particular kind of pitiless progress, and the skybound curves of its cables pointed heavenwards. The artist Joseph Stella painted Brooklyn Bridge this way in 1939; his image is almost robotic, a vision that mingles the bold primary colours of medieval stained glass with a cybernetic aesthetic that speaks of the coming age of technology. Seen from Brooklyn, the arches of the bridge's towers framed Manhattan as if through a cathedral window, as if the motion and energy of the city was more worthy of praise than any saint's deed or Calvary.

Chapter Nine

Bridge of Home

Forth, Scotland
Forth Railway Bridge: steel cantilever (1890), 2,467m
Border: Fife–Lothian

One feature especially delights me – the absence of all ornament. Any architectural detail borrowed from any style would have been out of place in such a work.

ALFRED WATERHOUSE, BRIDGE ENGINEER

For the first eighty years that the Forth Rail Bridge stood it was simply The Bridge. People who grow up in Fife or in Lothian feel pride in it, as well as a sense of ownership. A standing joke among Edinburgh folk maintains that until the Forth Bridge was completed, Fife was as good as an island, and was genetically inbred as a result.

It is still considered a wonder of engineering, the largest listed building in the British Isles and a UNESCO World Heritage site. The Ladybird book of bridges I read as a child lionised it – using it as a cover star, but also as a paradigm of strength, elegance and magnificence. It was completed in 1890 for £3 million (about half a billion today) – three great diamond webs of steel hammered out of eight million rivets and 50,000 tonnes of metal. The towers stand 370 feet high, its rail level at a more modest 156 feet. Its rust-red plates would stretch from Edinburgh to Glasgow if laid end to end. Each immense lattice rests on four concrete caissons, each one 70 feet in diameter.

I sailed around those caissons once in a dinghy, and the branching red network of columns extending skywards made it seem as if the sky had a capillary bed. The struts of the bridge have regularly spaced strengthening rings which, like the thickenings of bamboo, give the structure strength for a relatively modest weight. The columns reach towards one another in an expansive gesture. The Scottish biologist and mathematician D'Arcy Thompson wrote in his masterpiece *On Growth and Form* (1917) that its design 'embodied pure functional utility'. The textile designer William Morris called it 'the supremest specimen of all ugliness'.

It took seven years to build, and brought about the deaths of 57 men – with Victorian nonchalance, one of its engineers commented: 'It is, of course, impossible to carry out a gigantic work of this kind . . . without paying for it, not merely in money but in men's lives.' As its columns went up and its cantilevered arms stretched ever wider, boats were positioned beneath each pier to haul out any men who fell into the water. It's said that eight lives were saved this way.

Its hollow principal support beams are so broad you could drive a subway train through them – appropriately enough, given that its two principal engineers, Benjamin Baker and John Fowler, were involved in extending London's first underground railway. They also created the Nile dam at Aswan. One of the other engineers, Wilhelm Westhofen, used to climb to the top of the structure to admire the Forth Estuary laid out beneath him: 'The broad river itself, with craft of all sorts and sizes, in steam or under sail, cutting across the current on the tack, or lazily drifting with the tide, is always a most impressive spectacle upon which one can gaze for hours with an admiring and untiring eye.'

In a charity shop I picked up a copy of Westhofen's book about the bridge, and was startled to find pages illustrating a few alternative designs. One was a suspension bridge with bracing cable stays like the Brooklyn Bridge, but the estuary is so wide it would have required four towers; another supported itself through a series of crossed bracing chains. The diamond lattices of the chosen design, with its immense flying buttresses, are so much a part of my mental landscape that it was a shock to think that the bridge could so easily have been different.

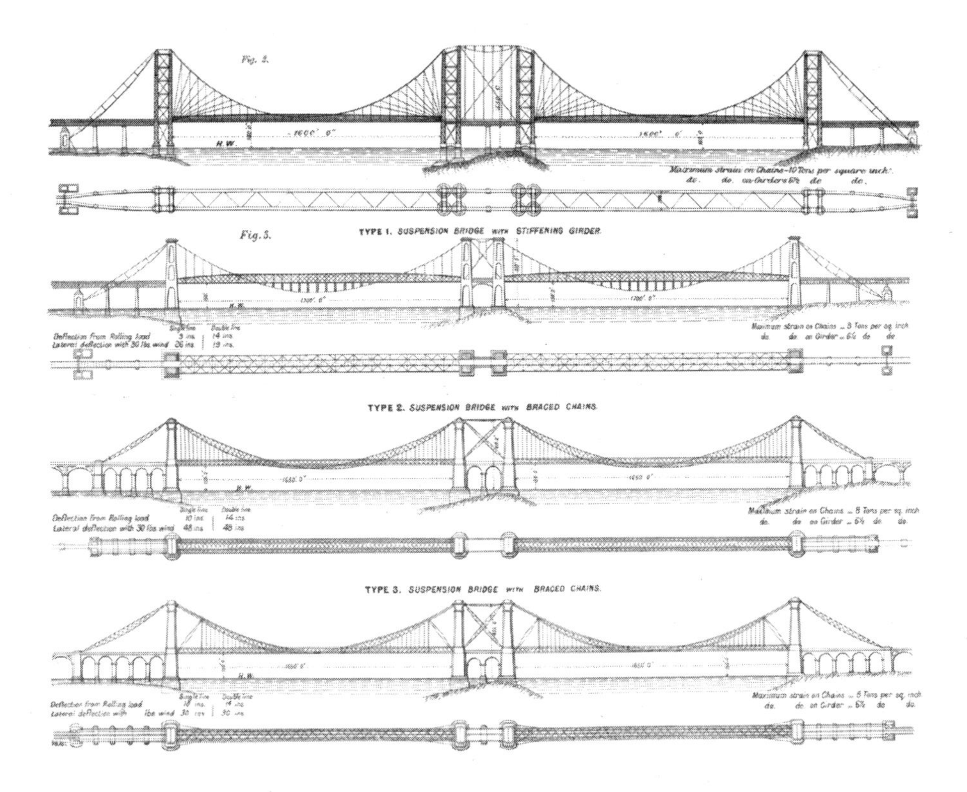

In the late 1990s, soon after finishing university, I took my first job as a doctor in the small general hospital of Dunfermline, in Fife. Most of my friends were working in Edinburgh, and my social life remained there, so I became familiar with the Forth Rail Bridge in a way I hadn't known it before, passing back and forth over it several times a week.

I'd been studying medicine for six years, but all those hours in the library hadn't prepared me for the dizzying, exhausting reality of work as a junior doctor in the UK's NHS, of gruelling rotas involving around a hundred hours on duty each week. On those trips between Fife and Lothian, I used to fantasise about what I'd do with the freedom of completing my junior doctor years – where I'd go once I had been liberated from exams, with money saved and the world open to being explored. After night shifts and morning ward rounds, I'd journey back to Edinburgh by rail; I'd always take a moment to glance down on the water as the firth widened to the east, into the sunrise over the North Sea, and think of how that sea was continuous with the Atlantic Ocean, which joined with other oceans in girdling the world.

There's a madcap but compelling theory that the coastline of Bruegel's painting *Landscape with the Fall of Icarus* depicts not some Mediterranean shore to which Icarus and Daedalus might have been expected to cross in their flight from Crete, but instead a view east down the Firth of Forth. The broadening of the Forth estuary to the east does resemble the kind of ideal (or idealised) landscape Bruegel portrayed. The features of his painting resolve into position with a nudge of faith: the island of Inchcolm with its two skerries, the mountain or 'law' of North Berwick in the haze of distance, the island of Inchkeith, the port of Burntisland to the north, the city of Edinburgh to the south, in the foothills of the Pentland Hills – all lie in the correct orientation to one another should Bruegel have

stood on a headland of the southern shores of Fife – just about at the point where the railway line emerges from cliffs of dolerite and stretches out across the water.

It's known that Bruegel did visit and paint the Forth estuary: the Antwerp traders he lived among frequented the Fife coast, and throughout the years he was painting they'd arrive in the estuary with their ship-holds full of roof tiles, and trade them for wool bales and barrels of salted fish. The villages of the Forth still use those red roof tiles as if in commemoration of their long-standing connections with the Low Countries, though they no longer sell salt fish, or pack so much wool. The painting is famous for the leg of Icarus flailing in the waves, and for the indifference of bystanders to his doom. It's an indifference commemorated by W. H. Auden in his poem about the painting, 'Musée des Beaux Arts' ('how everything turns away / Quite leisurely from the disaster'). For the Pulitzer Prize-winning poet and physician William Carlos Williams it was the painting's rendition of the panoply of life that was most striking ('the whole pageantry / of the year was / awake tingling').

When I told the art critic and writer Laura Cumming that I was writing a book about bridges, she told me that for many of the artists whose work she loves, bridges are places of transformation but also of ambivalence – often portrayed as heavenly, but just as frequently as hellish or dangerous. 'Bridges are as frightening as they are awesome,' she said. 'What immediately comes to mind is Munch, of course; and Monet, so those would be the hell and heaven bridges. I am profoundly shocked every time by the bridges in Daumier, whose figures get across them in fear and danger – there is even one of a pregnant woman hurrying across who doesn't make it. And I deeply love the many bridges of Hiroshige.'

For Cumming, what most appealed about Hiroshige's images were the ways they show human ingenuity and innovation pitted against the dangers of deep water, and the way the very delicacy and fragility of his prints seem part of their design, intended to echo the vulnerability and weakness of humanity as it hurries from one shore to another.

As a hospital intern there seemed no end to new procedures to be mastered, to the bleeps from the nurses, to the pages of blood results to be checked, to the flow of broken people, and I began to yearn for the road, for travel – there were still so many places I was yet to visit. On the train back towards the hospital, after nights out with friends, I was usually so fatigued that I'd fall asleep; if I dreamed, those dreams would be of flight.

2000s

Chapter Ten

Bridge of Vertigo

Golden Gate, USA
Golden Gate Bridge: suspension (1937), 2,737m
Borders: San Francisco Bay–Pacific; life–death

The Golden Gate Bridge is practically suicide proof. Suicide from the bridge is neither possible nor probable.

JOSEPH STRAUSS, CHIEF ENGINEER, GOLDEN GATE BRIDGE

Twenty minutes into Eric Steel's film *The Bridge*, a documentary examining the phenomenon of suicide from the Golden Gate Bridge in San Francisco, a mother and daughter are interviewed. Lisa, the youngest member of the family, who had a diagnosis of paranoid schizophrenia, had jumped from the bridge after years of mental health difficulties. 'I think it was a relief for her,' says her mother. 'The way I look at it she's in a better place.' 'You have to look at it that way,' adds Lisa's sister. Four minutes later, a father is speaking of his son's jump from the bridge: 'He thought his body was a prison – his mind – he knew he was loved, he knew he had everything, could do anything, and yet, he felt trapped. And that was the only way he could get free.'

In his essay *Notes on Suicide* the philosopher Simon Critchley points out that there can sometimes be a narcissistic edge to suicide – an edge that's sharpened by such a dramatic choice of exit as the Golden Gate Bridge. 'As everyone knows, the Golden Gate Bridge is a popular suicide destination,' he writes. 'Yet all the suicides jump from the side of the bridge that faces San Francisco. No one wants to jump from the side that faces out to the Pacific Ocean.' That helped Steel's project, whose cameras caught jumper after jumper over the course of a year, some teetering out over the edge and almost stumbling off, some leaping with certainty, their legs pedalling as they fell, and one victim treated as a kind of grotesque teaser trailer through the latter part of the film, shown at the end tumbling from the rail, arms outstretched, as if the bridge was his scaffold or gallows.

Just under an hour in, Steel puts the camera on a young man who was asked by a tourist, just as he was preparing to jump from the bridge, if he'd take a snapshot. The young man took the photo for the stranger (as human beings, what do we owe each other?) then walked to the edge and jumped. But as soon as his feet left the metal girder he knew he'd made a mistake. It takes between four and seven seconds to reach the water from the Golden Gate Bridge, depending on the way you fall; he had the presence of mind to angle his body and tuck in his arms so that his feet hit the water cleanly. Two of his vertebrae were shattered, but he survived (and believes a seal saved his life, by nosing him back to the surface). Many of those who jump must survive the initial impact only to drown afterwards.

Steel's documentary has many critics; it's as if he has looked through a journalist's list of 'don'ts' when writing about suicide, and set out to 'do' every one of them. But watching it can be a redemptive experience: family after family confess their anger and frustration over what happened, their irritation and regret, but they also describe their love for the one who jumped, and their gratitude that such suffering has come to an end. One friend berates herself for what she might have done differently, but is followed by another who reminds us how grey is the zone between doing little to prevent a friend's suicide, and attempting to do everything. One interviewee feels he could have had his suicidal friend 'locked up or something' to save him, but recognises that to think like that is to invite torment: 'I don't blame myself like that,' he says. Watching that interview I found myself nodding. If I blamed myself 'like that' for the suicides I've known in thirty years of medical practice, I don't know if I could go on.

One Tuesday afternoon a couple of years ago I arrived at the old Forth Road Bridge too late to prevent a suicide. I was crossing by bicycle when two desperate women with flailing arms flagged me down just as I approached the south tower. A man had jumped, they told me, and they didn't know what to do. 'How long ago?' I asked. 'Seconds,' they replied, 'just now.' I rushed to the railing and looked down on the grey waves to see a man's bare back, red shorts, blue running shoes. The impact must have ripped off his T-shirt. He lay face down, his body rippling with each wave, slipping upriver with the rising tide.

He had marched to the barrier, they said, pushing his earphones deeper into his ears as he had passed them. 'A big man,' one of the women said, 'big like this,' and she squared her arms out from her sides. She had thought of asking him to take a picture of the two of them on their happy day out, the magnificence of the estuary and its rail bridge behind them, but on catching a glimpse of his expression had changed her mind. The man had reached the tower only a few yards ahead of them, and climbed over the railing. She'd shrieked and run towards him, but after a single glance in her direction, he had jumped.

The woman was hyperventilating, holding her phone out to me and asking who we should call. I laid what I hoped was a comforting hand on her shoulder as I called the police and coastguard from her phone. Then I got on my bicycle and circled back off the bridge and down to the shoreline. I hoped he had drifted inshore enough that I might yet be able to reach him, maybe even haul him out of the water and see if he was still breathing, but on my arrival under the decks of the bridge I could see that the tide was pulling him further away from the beach. I felt helpless, almost panicked, could feel my heart pounding; it had now been almost ten minutes and I hadn't seen him lift his head. I was sure now that he must be dead.

After only a few moments two police officers arrived, jogging to where I stood at the water's edge, immediately beneath the bridge, and I guided their grey, searching eyes to where his body was floating out, further still. They in turn, with windmill gestures, began to guide a coastguard launch with a crew of four men who, in their orange life vests, managed with difficulty to pull the body from the waves. Their launch turned, and sped back in the direction of the lifeboat pier.

Those two police officers kept their gaze grimly towards the ground as they strode back in the direction of their vehicle, job done. I wondered how many times a year they answered calls like this – only the week prior I'd heard the blades of a search helicopter over the water – but wished all the same for the comfort of some shared gesture. In an earlier age we would have crossed ourselves, perhaps – something to acknowledge and honour the sudden and violent passing of a life. 'Do you need a statement from me?' I asked as they receded, raising my voice against the wind. But they only shook their heads with downturned mouths, got into their patrol car and drove away. I pedalled back up to the deck of the bridge, but it was empty. The two women too had gone on their way. The waves beneath me rolled on as if none of this had happened.

In the year 2000 I moved on from being a junior doctor at a general hospital in Fife to begin training in emergency medicine at the largest hospital in Edinburgh. It was the year Google began selling advertising linked to web searches; the dot-com bubble was inflating. The Bay Area of San Francisco, home to so many of the world's richest companies, was on its way to having the most expensive real estate in the United States. I had just been told about a new way of staying in touch with friends – 'hotmail'.

One of my first patients as an emergency physician in the city hospital was a man who had jumped from the Forth Road Bridge and survived. He had shattered his ankles and three of his vertebrae in the fall. To land on water from a height of 150 feet is like falling on solid ground. The psychiatrist told me that no one jumps from a height like that on a whim, and so the man would need very careful observation and support. Neither was it the first time he had tried to end his life. When visiting time came I remember his family gathering around the bed, their hands trembling and their faces anxious as they struggled to find words to express their sadness and relief.

As a doctor learning the trade of emergency medicine it often fell to me to break bad news to the families of people who'd reached A&E too late, or too broken, to survive. Often the bereaved families had sat in the little family room many times before, on other occasions when their brother or mother, sister or spouse had jumped or taken poison. There was horror, shock, grief and tears, but occasionally I also detected something akin to, but at the same time different from, relief. I admired the way some of those bereaved families managed to find consolation in such an appalling end for their loved ones – in that, with that final act, a great suffering had come to an end.

Suicide, falling and the Golden Gate Bridge are the recurrent themes of Hitchcock's film *Vertigo*, voted every so often the greatest movie of all time. I watched it for the first time that year, in those weeks when I was beginning to encounter suicidal people as they were carried into the emergency room, shattered, bleeding or sluggish with overdose.

There's a scene in the film where Madeleine (Kim Novak) is pacing back and forth beneath a shoreside span of the great red bridge over the Golden Gate. The clouds shadow one another reciprocally like smudges of silver nitrate; between them, the sky is the blue of delicate veins at the wrist. The immense towers of the bridge dominate the frame like Californian redwoods;

trusses underpin the decks and make a scaffold over Madeleine's head. She seems agitated and unsettled, then suddenly she leaps into that great upwelling of abyssal water that laps America's western coast, water as grey and rough as a shroud. The man hired to follow her, an ex-detective called Scottie (James Stewart), follows her into the water, supporting her as she lies surrounded by the flowers of her bouquet like a pre-Raphaelite heroine. For a moment their two heads bob in the water as if both have drowned.

Of all the visual symbols in the film – spirals and flowers and redwoods and towers – it's the uprights of the Golden Gate Bridge that are the most memorable. Hart Crane wrote of the highways of America between New York and the Golden Gate as a great bridge of tarmac:

> Macadam, gun-grey as the tunny's belt,
> Leaps from Far Rockaway to Golden Gate:
> Listen! the miles a hurdy-gurdy grinds –
> Down gold arpeggios mile on mile unwinds.

Its towers stand like a semi-colon at the end of America, there where the sun goes down on the New World. The setting of the Golden Gate Bridge creates a dramatic harmony of sea and sun, light and dark, death and renewal, even of eternity and oblivion.

American writer Rebecca Solnit has described *Vertigo* as a love letter to San Francisco. For Solnit, the film's most interesting character is Scottie's friend and ex-fiancée Midge – the rhyme with bridge seems deliberate – who is a designer of underwear. Midge broke off her engagement to Scottie – we never hear why, though some critics have pointed to hints at Scottie's impotence – but she adopts a maternal care for him that shades into regret when he begins to fall instead for Madeleine.

At the beginning of the film Scottie pays Midge a visit, complaining of pain from a corset he wears to keep his spine straight, following an injury he sustained in pursuit of a thief. During that chase, a policeman died, and now Scottie suffers from acrophobia, a fear of heights, which gives him the dizzying and incapacitating sensation of vertigo. We never see how he was

rescued from his position dangling from the gutter of a house, and so from the opening scenes of the film he is in a way left dangling in the viewer's mind over a great chasm, death waiting for him the way it awaits all of us, as he hopes for the redemption of a ladder, a rope or a makeshift bridge.

The Golden Gate was the longest suspension bridge in the world at the time of filming in 1957 (work on long suspension bridges worldwide had halted after the Tacoma Narrows bridge disaster of 1940, in which a too-flimsy suspension bridge had begun to undulate in a crosswind and broke into pieces). The Golden Gate Bridge was by then a 20-year-old colossus of an art deco sculpture, and it bears the stamp of its decade. It wouldn't be surpassed in length until the Forth Road Bridge, a far plainer, more timeless style, in 1964 (work began on the Forth Road Bridge the same year *Vertigo* was released). The two have something else in common: they are both bridges notorious for suicides.

* * *

After a couple of years' training in emergency medicine, my confidence as a physician was growing. I was beginning to think about how to build a life and a career that would allow me to practise the profession I loved, but would also allow me to travel and see more of the world, and make connections with more (and different) kinds of people.

My work today is no longer in hospital emergency medicine, but in community clinics as a general practitioner. In the course of that work I regularly hear people tell me they wish they were dead, or that they are planning their suicide. The patients who pass through my clinic asking to be cured of their suicidal thoughts are as diverse as our communities: unhappily married spouses, harassed single parents, teenagers, isolated pensioners, adult survivors of childhood abuse. I was taught by the psychiatrists at medical school to gently swing these conversations round to explore the social connections that sustain life, and then explicitly document them. These bridges between the self and others are called 'protective factors' in the spare language of the clinic. I was to find a way to remind my patients that every suicide punches a ragged hole through our collective social fabric, and that such holes are not easily mended.

A therapist told me once about a psychotherapeutic technique called the rope bridge metaphor, which she would use with her patients. It reminded me of the message hidden within the story of the Billy Goats Gruff: a client is encouraged to imagine the life they would like to have as terrain on the far side of a canyon, or across a fast-flowing river. On *this* side of the canyon they're stuck with the life they have now; on the far side there's the promise of a new life. Connecting the two is a spindly rope bridge with wooden slats that look as if they are falling through. To reach that imagined life is a risk, and there will be many who baulk at such a crossing: the bridge seems too fragile. For others it will be clear that there is some work to be done, changes to be put in place, achievable preliminary repairs, before taking that first but necessary step.

But sometimes counselling and finding ways to talk about your emotional and psychological pain doesn't help – some of the suicidal patients I've known have been tipped into crisis by revisiting traumatic events in their past. There are two ways of dealing with profound emotional trauma: you can build a bridge to it, or you can cut off all connection to the memory and leave it undisturbed. I've known some unfortunate patients over the years who have found therapy profoundly *un*helpful: occasionally to build a bridge to a place of pain and trauma is to open yourself to destruction.

The suicides in *Vertigo* are all fake – arranged to dupe Scottie, and the police, into believing the death of another young woman (the wife of Madeleine's lover Gavin) was suicide rather than murder. Scottie has been set up, because Gavin knows about his fear of heights. The Californian landscape scenes of the film, the most climate-blessed of the States, are revealed to be the backdrop not to repeated attempts at suicide, but attempts to cover up a murder. The symbolic power of the bridge as the gateway not to the Pacific but between life and death isn't diminished by the revelation. It looms behind and above the characters, ominous, like the madness of despair that leads people to suicide.

During the opening scenes of the film Midge shows Scottie a prototype 'uplift' brassiere which works on the cantilever principle, using a series of struts and balances to defy gravity – a bra that obliges breasts to support themselves, without back or shoulder straps; a bra that stays up while Scottie's terror of falling keeps him down. Her studio is a workspace of book-covered desks, easels, diagrams and models; she's understood as a self-made woman who has made a safe haven for herself, with a view out over all the turbulence and dynamism of San Francisco, its houses packed tightly in whorled, haphazard streets. Rebecca Solnit wrote of Midge: '*Vertigo* is a film with a paper doll Tristan and Iseult sliding across the foreground and this round figure making one startling appearance.'

Madeleine by contrast is a cipher, insubstantial, and when we eventually find out where she lives, it's an anonymous hotel room. '[Midge] is an invitation to go in another direction than the tragic one of the film, for though the movie is in love with San Francisco, she is the only character who really seems immersed in the city's possibilities.' Unlike Scottie and Madeleine she isn't tangled up in fantasies of death and lust, but stands as the cantilevered bridge back to sanity that Scottie doesn't see; his unfulfilled redemption. She's a reminder of the consolations of life – of pleasure, love, and pride in one's work – and of what good foundations and careful planning can sometimes achieve.

Chapter Eleven

Bridge of Defence

Urubamba, Peru
Several rope bridges of the Incas (1420)
Border: Spanish–Quechua culture

It was destroyed by Spanish conquest, and the world will never see its like again. A few of the destroyers, only a very few, could appreciate the fabric they had pulled down, its beauty and symmetry, and its perfect adaptation to its environment. But no one could rebuild it.

CLEMENTS MARKHAM, *THE INCAS OF PERU*

There's a film made in the 1930s, somewhere in the basin of the Congo River, that shows a group of indigenous forest people building a suspension bridge across a tributary of the river. It's jungle; high trees lean in over the water from each side, and the group's best climber is sent to the top of the tallest. He attaches a long rope made of vine, fitted with a saddle, and drops it. A young boy then climbs into the saddle and, pulled back by a team of men, swings out over the water where crocodiles bask. Higher and higher he swings, great arcs of motion over the fast-flowing river, until with a hooked stick he catches hold of a branch on the far side. From his vine-tied cradle he climbs out into the flimsy branches of the canopy and makes fast his rope, then crawls back across it like a spider along a silken thread. With one section of the bridge anchored he climbs back and forward bringing more ropes, more reinforcement, until multiple lengths of vine cross the river. A team of weavers now move in, climbing up with their coiled reeds and grasses on wooden gantries set leaning against the trees. They weave a thick mat of fibre, a deck that even the youngest of children can be carried across. The bridge has taken eight days to build.

In the ancient language of the Quechua people Cuzco means 'umbilicus', as in, the centre of the world. The city sprawls across a shallow depression in the mountains, seeping into countless tiny valleys and clambering up into the edges of the great Andean forest. Its terracotta roof tiles seem more Mediterranean than Andean, its basilicas, cupolas and baroque facades more Florentine than Pan-American. Vast colonial plazas built on the wreckage of Inca temples are lined with ornate and triumphant churches. The stones of the old Inca walls fit together as neatly as atoms in a crystal, standing firm after hundreds of years through the earthquakes – though the Spanish-inspired architecture crumbles a little more at every tremor.

When the Spanish first arrived the central plaza was twice the size it is now, the temple walls were encased in gold, and there were buildings dedicated to the moon and stars, each of which brimmed with silver. The conquistadores were more at home in the gambling dens of Toledo than on the brink of a new world, but within a matter of months, a few hundred of them had subjugated the warrior race who only a few decades before had conquered South America from Patagonia to Colombia.

The paintings of the Cusqueña school of religious iconography hang in every church in the city, and they always seem to portray Jesus's Roman jailers in the armour of the conquistadores. As the 'Incas' (though 'Inca' referred specifically to the king) were beaten back by the Spanish, they retreated further and further into the mountains, across rope bridges, into hidden fortresses built high over the Urubamba and Vilcanota rivers.

The valleys are narrow and steep, and when I visited in 2002 the rivers were engorged with summer rains, loaded with silt like molten chocolate. Tiny ripples on the water were compact and elegant, tildes on a Spanish ñ. In some places the valley broadens to a flood plain, and graceful curving terraces step away from a stitched patchwork of fields in every shade of green, their width shortening as they gather gradient until they are as broad as they are high; the mountain carved into the contours of a topographic map.

The Spanish followed the Quechua down these valleys but didn't find all their hiding places. The slopes are so precipitous and the forest so dense that Hiram Bingham, an American archaeologist, only 'discovered' Machu Picchu in 1911. It was a capital high in the mountains reachable by spindly rope bridges. The Quechua knew about military fortifications, and they knew how to hide from aggressors.

The ingenuity of the Quechua's hidden world is astonishing – rope bridges that could be raised or dismantled with ease, and on one rocky approach to the city, the path traversed a shelf carved into the rock. Part of the ledge was cut away, and could only be crossed by a single timber plank. To seal off the city – to protect it from the damaging influence of traffic and exchange – it was enough to raise this plank, cut down a rope bridge, and the city was safe.

I dutifully stood in line and signed up for one of the guided tours that walk the old Inca highway to Machu Picchu, a journey of three and a half days. Until recently, I was told, it had been possible to hike the trail on your own, but the practice of walking it without a guide had been stopped because it was having unpleasant side effects: an unwholesome accumulation of both sewage and Peruvian bandits.

The trail began with a rope-bridge crossing of the Urubamba River, swaying out over the water beside an older, rotten model whose planks had mostly fallen in. The fear of bridges has an unlovely name, gephyrophobia, and it's a fear I never understood until I stepped for the first time onto a frail and bouncing rope bridge. But the experience was also vital and invigorating in an unexpected way.

I stepped from Plank to Plank
A slow and cautious way
The Stars about my Head I felt
About my Feet the Sea –

I knew not but the next
Would be my final inch –
This gave me that precarious Gait
Some call Experience

Emily Dickinson's poem of traversing a pier or a bridge between the stars and sea is far more than a meditation on fear and knowledge; it suggests the transformations of perspective that can be offered by bridges both actual and metaphorical. Even the infamous dashes '– –' with which she peppered her poems stand like little bridges between the reality of her words and the numinous reach of her intentions. Words can be thought of as bridges between minds, and between the realities of the world and our mental abstractions of it. Partway over a bridge crossing, the world behind you has shrunk, and the new world on the far bank is not yet real. As you ascend (or in the case of a rope bridge, descend) the arch of a bridge, you too take up less space within that world, and begin to imagine your own absence. Is the love of crossing bridges, gephyrophilia, akin to the appeal of mountaineering? To gather new perspective on the space you take up in the world, and be half in love with your own insignificance?

My journey through Peru, Bolivia, Argentina and Chile was to take advantage of a few months' gap in my working schedule – between quitting my job in emergency medicine and taking up another job as an expedition doctor. I'd never been so far from home before, and part of me was preparing myself for my next posting – eighteen months in Antarctica, the wildest, emptiest, most elemental wilderness on earth. It was as if my travels were going to take me a very long way down a narrow, blind-ended bridge. Latin America seemed like a good place to test the elasticity of my connections with home – with my family and all those people dear to me. It was as if I needed to see how it felt to be on the other side of the planet before moving to a place so isolated that it would be easier to be evacuated from the International Space Station than from my new home in winter.

Beyond the Urubamba rope bridge the path climbed up out of the valley to 4,200 metres before falling sheer into the jungle, then winding back up to Machu Picchu itself. Rough-hewn slopes of green dropped into narrow gorges, and clouds drifted in and out of the valleys like thoughts. There were occasional rainbows – in Quechua tradition, bridges between heaven and earth. I thought about how, for the Greeks, rainbows were not so much bridges between heaven and earth as the multicoloured trail of Iris as she streaked across the sky carrying messages from the gods. Orchids hung from the branches overhead; they were delicate and golden like curls of butter. Coca leaves as shiny as polished leather grew by the pathside, and I chewed them until my tongue and lips were numb.

On the fourth day we walked for a couple of hours in darkness so as to arrive at Machu Picchu at sunrise, every sense overwhelmed except that of sight. The air was thick with dank humidity and laden with a million organic smells, of the jungle growing, dying and rotting. Each tree was a library of hidden knowledge. Our surroundings reverberated with sound: of birds shrieking alarm, and frogs creaking among the vines and moss. It was delicious in part because of my awareness that soon I'd be in Antarctica – a place without plants, without smells, a place of light, ice and silence. When we arrived at the city it was in mist, and it wasn't until midday had come and gone that the sky cleared, the sun came out and the ruins revealed themselves.

> The sanctuary was lost for centuries because this ridge is in the most inaccessible corner of the most inaccessible section of the central Andes. No other part of the highlands of Peru is better defended by natural bulwarks – a stupendous canyon whose rock is granite, and whose precipices are frequently a thousand feet sheer, presenting difficulties which daunt the most ambitious modern mountain climbers. Yet, here, in a remote part of the canyon, on this narrow ridge flanked by tremendous precipices, a highly civilised people, artistic, inventive, well-organised, and capable of sustained endeavour, at some time in the distant past built themselves a sanctuary for the worship of the sun.
>
> HIRAM BINGHAM

The stonework of the Quechua people seemed almost miraculous in its artistry; caves fashioned into condor wings, and tiny aqueducts carved into walls that still trickled rivulets of mercurial water. On top of the central hill stood a cubist altar used in the ritual disembowelment of llamas (a bit had been chipped off, I was told, by the heavy equipment of a television crew making a commercial).

The little I knew of the Quechua world was bewildering to me, and to the ways of thinking I'd grown up with, utterly alien – involving as it did human sacrifices, bodies broken and stuffed into pottery jars, left on Andean peaks as offerings to a sky god or Mother Earth. These bridges of the high Andes were, more than any others I'd yet seen, bridges between worlds.

In my backpack I carried a short novel written about the rope bridges of Peru. *The Bridge of San Luis Rey* by Thornton Wilder won a Pulitzer Prize for its examination of one friar's search for meaning in life and death, after the collapse of a Peruvian rope bridge caused the deaths of five innocent people. Wilder fictionalised the precarious rope bridges of Peru as a metaphor for the precarity and preciousness of life – his 'Brother Juniper' spends years seeking the backstories of each victim of the bridge disaster, trying to make sense of their ultimately senseless deaths: an orphan brought up in a convent, a noblewoman, a streetwise theatre impresario, the noblewoman's maid, and the son of a famous actress.

The friar fails in his mission to connect the five deaths to any purpose, and is rewarded for his spiritual search by being burnt at the stake for his audacity in questioning God's plan. Meanwhile the relatives and friends of the victims seek solace in one another, and in the building of community. The novel's final lines cite the bridge as the ultimate metaphor for traffic between spiritual worlds. The connections we build with other human beings are framed as the sole consolation for the fragility and suffering of being alive. 'Soon we shall die and all memory of those five will have left the earth, and we ourselves shall be loved for a while and forgotten,' Wilder wrote. 'But the love will have been enough . . . There is a land of the living and a land of the dead and the bridge is love, the only survival, the only meaning.'

Two years into speciality training, my peers had fanned out across the world, to take up specialist training programmes in London and Sydney, Harvard and Oxford. It seemed as if they were sinking with certainty into their careers while I, with my uncertainty, was unwilling to abandon the possibilities of travel and the experience of freedom. My own path seemed much more precarious.

Shortly after leaving Latin America I embarked as a ship's doctor at a wharf in Humberside, then sailed south for many weeks, the length of the Atlantic. It seemed for those weeks as if Emily Dickinson's poem was not a description of crossing a swaying rope bridge set with planks, but of moving along the wooden deck of a ship: *I stepped from Plank to Plank / A slow and cautious way / The Stars about my Head I felt / About my Feet the Sea –*

In Antarctica I'd test my capacity for isolation, but also my connections with the people I loved. I understood Dickinson's image as one of tiptoeing on a precarious journey, in search of extreme experience. Out on deck each night, the ocean glittering with reflected starlight, the rise and fall of the ship like a bouncing rope bridge.

Chapter Twelve

Bridge of History

Hellespont / Çanakkale Boğazı, Türkiye
Xerxes' bridge: pontoon (480 BCE), approx. 2,500m
1915 Çanakkale Bridge: suspension (2022), 2,023m
Border: Europe–Asia

The Hellenic race . . . continues free, and is the best-governed of any people, and, if it could be formed into one state, would be able to rule the world.

ARISTOTLE, *POLITICS*

Approaching Istanbul, the 'Queen of Cities', from the north there's an uncanny feeling of being at one of the world's greatest gathering points of geography: land and sea, history and culture, north and south, east and west, they all meet here. Yet through all this flux and transition Istanbul feels timeless. To watch supertankers glide through one of the world's megacities seems slightly unreal, and that's before one begins to contemplate that this is the same water the Argonauts are said to have rowed on their way to claim the Golden Fleece, that the Vikings defended as Byzantine mercenaries, that the Venetians fought the Genoese over, that Mehmed the Conqueror besieged, that gave onto the seas of the nineteenth-century Crimean War. The Vikings called Istanbul simply Micklagard – the Great City. It was restricted to a western peninsula of the European shore back then, beside the inlet of the Golden Horn (from which the Golden Gate in San Francisco takes its name), but now sprawls for tens of miles on both sides of the strait. From a population of 500,000 at the declaration of the Turkish Republic in 1923, the city has swelled to 12 million inhabitants.

A year after returning from Antarctica I married, and a year after that my new wife E. and I resolved to travel around the world on a motorcycle. To reach Istanbul we had driven down through Western Europe to Greece, then passed over the Evros River at Greece's border with Türkiye. We had hardly noticed the bridge that constituted the Greece–Türkiye border – there was so much concrete, so many billboards and barriers, that I passed over its water oblivious. The landscape was the same on the other side, but horse-drawn carriages became more frequent, the women at the roadside wore hijab, the villages had mosques instead of churches. Many Turks have worked in Germany, so although I found myself mute across much of Greece, with my high-school German I could make myself understood. To the Romans this was the Via Egnatia, a military road since long before the invasions of the Persians, and even today it is broad and smooth, robust enough to support tanks and heavy artillery. Greece and Türkiye have for several years matched their military spending at around 3 per cent of GDP – for the latter that's USD $20 billion – a much higher level than comparable countries around them. The border bristled with barbed wire.

Istanbul's suburbs began with rows of dirty white concrete blocks, batches of identically shaped and painted apartments crusting the hills like barnacles. The road carved through them, the traffic becoming denser as we approached Taksim Square, at the city's commercial centre. The sunset call to prayer, the car horns, the hubbub of crowded streets all mingled into a beautiful cacophony, the city like a fevered and delirious poem.

As I grew accustomed to Istanbul my eye was forever being drawn to the agitated water of the Bosporus, in motion even on days that were otherwise grey and still. Also changing was the light over the Aya Sophia – the principal cathedral of Christendom for a millennium. Some ferries dropped gangways like the jaws of Jonah's whale; others simply bumped against rubber tyres slung to the sides of the piers, while commuters between continents jumped on or off before the ferry could depart again. Above my head, over three gigantic bridges, almost half a million people crossed the water every day.

We crossed the Galata Bridge over the Golden Horn into the old Byzantine city, edging past its rows of men holding fishing rods. The bridge had only been built in 1845, though Sultan Bayezid II had, in the early 1500s, encouraged both Michelangelo and Leonardo da Vinci to submit designs for one. Leonardo's was a magnificent shallow arch that, had it been built, would have brought forward European bridge engineering by centuries. Looking at diagrams of it online I was reminded of Leonardo's depictions of human anatomy that also remained unpublished. Had his bridges and his anatomies been more widely known, he would have revolutionised both medicine *and* engineering.

I never tired of watching the light on the water change with the pageantry of the wind and clouds, the way it dappled the strait that divides Europe from Asia. But that divide was here irrelevant, dissolved in the elision of ferry and bridge, culturally identical on both shores. The main Bosporus Bridge itself might have made me homesick for the Forth Valley if I hadn't been so in love with the city beneath it. It shares designers with the Forth Road Bridge – both built by the international engineering company Freeman Fox & Partners (as was the Zambezi bridge in southern Africa). But the bridge over the Bosporus stands higher than the Forth, its cables engineered to take a deck that's broader in width but shorter in length. It towers over the city like a piece of furniture made for a Titan, one that plants a foot in Europe and a foot in Asia.

After a fortnight in the city we drove the motorcycle across that bridge and into Asia. Out on the steppe we stopped at a museum built to celebrate the tomb of King Midas, and which commemorated Alexander the Great and his cutting of the Gordian Knot. The custodian of the tomb had a small moustache; he took a fresh fig that I offered, and unlocked the gates for us. In such museums the warring cultures that have made their home here – Roman, Greek, Persian, Ottoman, Hittite, Mongol and the rest – are finally reconciled. It was small and cool in the shade, with quiet displays of pottery and statuary. The ruler who could untie the Gordian Knot was said to be destined to rule over all Asia, which was why Alexander the Great was said to have sliced it apart with one stroke of his sword.

The next Anatolian museum we visited, in the former Byzantine capital of Iznik, was housed in the semi-ruin of an old mosque; I was shown Neolithic skeletons in pots jumbled alongside Hittite carvings. Sunlight fell through its windows onto marble reliefs of the labours of Hercules. Tombstones of Greeks, Romans and Ottomans were jumbled together in piles, evidence that the borders that have been erected around this land never last for long, and that the many peoples who have warred over them are united in death whether they like it or not. It's always been a rich land: a sarcophagus for a Roman merchant there was more magnificent than those sculpted elsewhere for kings. There were reliefs of atrocious battles, but also a delicate marble statue of Cupid. I watched an old Turkish man circle it with fascination, smoking as he gazed, as if trying to make sense of something that could never be resolved.

Another few hours driving south along the Sea of Marmara and we reached the Hellespont – or Çanakkale Boğazı as it's now known. The waters that divide the continents here glittered like a knife unsheathed, held in the teeth of the Gallipoli hills that stand opposite the archaeological site of ancient Troy. Our arrival coincided with 18 March, Turkish Victory Day, and the streets clinked with military brass on parade. There was a festive atmosphere: triumphal music played from loudspeakers hung from the buildings, and flags and bunting were strung between the lampposts.

There have been nine citadels excavated at Troy – its modern name, Hisarlık, means 'the place of fortresses'. The city seems to have been a strategic and commercial hub throughout antiquity, controlling access to the Sea of Marmara and the Black Sea. The Troy of Homer is thought to be the seventh of those built on the site, a city that flourished over three thousand years ago. Archaeologists have found signs of it having been scorched by fire then abandoned, just as Troy must have been.

The Scamander River has been dumping silt since the events of *The Iliad*, and what must once have been a protected harbour with marshy shores is now an open fertile plain. If we accept Homer's testimony, that plain hides the graves of unnumbered Greek soldiers, their blood now part of the soil of Asia. 'I could not name or even count them,' the epic recounts, 'not if I had ten tongues, ten mouths, a voice that could not tire, a heart of bronze.'

For the ancient Greeks the habitable earth was surrounded by Oceanus, the world river, a gulf that would be bridged only in death. We parked the motorbike on a hill facing west, onto the Aegean. The sea shimmered in a broad arc along the western horizon, embracing the land in light; the sky was a polished breastplate. The Scamander valley seemed sculpted as if to form a superb amphitheatre, the backdrop for a theatrical climax. But there was little drama now – a tractor ploughed the earth, and I watched it raise clouds of the same glinting dust that once plumed from the wheels of Greek and Trojan chariots. The place was quiet after the last tour bus had left for the day. The smells on the air were of wild rosemary and thyme, the smoke of burning olive branches, the earthy mineral smell of rocks cooling after the warmth of the afternoon. It was a restful place. I wondered whether the conflicts of history are repeated because of something in the nature of mankind, or in the nature of the gods that we let rule over us.

'Hellespont' means 'Sea of Helle', and takes its name from a mythical girl who was being spirited from Asia to Europe on the back of a flying ram. Pindar and Aeschylus tell the story, which hints at the dangers of traffic between the continents – she fell into its waters and drowned. This strait that divides Europe and Asia was thought by the ancients to represent a division between worlds, but the reality is more prosaic. It is a 'drowned fault valley', a crack in a continental plate that was flooded only within the last few thousand years. There's a theory that its inundation gave rise to the story of the Deluge – of Noah and his Ark. The city of Troy was built to defend the approaches to the Black Sea from the Mediterranean: to defend Asia from Europe.

The oldest bridge in the world still in use is in Türkiye. The Kervan Köprüsü at Izmir is thought to have been built around 850 BCE, at the terminus of the Persian highway across Anatolia, which carried traffic then, as now, from Asia Minor down to that coast's principal port.

To reach Europe without a sea passage, a different kind of bridge was built a couple of centuries later. The Hellespont was first bridged during the Graeco-Persian wars by a Persian emperor who wanted to create a strong state that straddled Europe and Asia. In 2022, it was bridged again, by a resurgent Turkish state whose ambition was the same. Persia or Türkiye, Xerxes or Erdoğan, the bridges of the Hellespont symbolise mankind's attempts to get around the barriers nature puts before us, as well as the forces by which peoples are kept divided, or are united.

Herodotus, the 'Father of History', ridiculed the idea that Europe and

Asia were condemned to eternal division; what was needed was simply a little more understanding on both sides. According to him the Trojan War, and the battles that succeeded it, had more to do with the practice of bride-stealing than irreconcilable differences between peoples. Invoking Helen of Troy he wrote: 'To abduct women is considered the action of scoundrels, but to worry about abducted women is the reaction of fools.' The first page of his famous *Histories* – itself an attempt to explore the roots of Europe's divisions with Asia – is dedicated to explaining the tit-for-tat theft of women that had gone on between Europeans and Asians for centuries. *The Iliad* of Homer and the *Histories* of Herodotus could be described as the twinned foundation stones of the literature of Europe, though both books have their origins in Asia.

Herodotus is our source for a description of the first Hellespont bridge, built in 480 BCE to facilitate a Persian invasion of Greece. It was a 'pontoon' bridge of boats anchored and lashed together. The gods were then believed to determine success or failure in human affairs, as well as to animate the non-human world. When the first attempt at building a bridge failed, destroyed by a storm, the Persian king Xerxes didn't just order its engineers beheaded, he had iron fetters forged and thrown into the waters, then demanded that the strait itself be whipped.

On the Persians' second attempt the weather was kinder, and their army was able to cross over to Gallipoli, and thus to Greece. 'Xerxes made libation from a golden cup into the sea,' wrote Herodotus of that first crossing of the bridge, 'and prayed to the Sun, that no accident might befall him such as should cause him to cease from subduing Europe, until he had come to its furthest limits.' *Subdue Europe unto its furthest limits.* The armies of Persia never reached much further west than Athens.

From the shoreline below Troy E. and I watched many ships sailing through the strait, among them several Russian and Ukrainian vessels. Their grey bulk flowed between Asia and European Gallipoli in a great stream of international commerce, wheat from Ukraine, gas from Russia. We're hearing more in the world today about the necessity of building walls and strengthening frontiers, even between nations that have traditionally been at peace, or even joined in union. When Russia annexed Crimea in 2014, work began on an immense bridge to connect Asia to Europe, from Taman in mainland Russia to Kerch in Crimea – a bridge to consolidate and extend the power of the Russian empire. Its 19 kilometres of twinned road and rail carriageways was completed in 2020. Like Xerxes' bridge, it became an artery supplying heavy military equipment from east to west, in order to wage war. In 2022 it was destroyed, probably by Ukrainian special forces, who ignited a truck full of explosives partway across the span. It has since been repaired.

On 18 March 1915 Winston Churchill ordered an Allied naval assault on the Hellespont in an attempt to drive Ottoman Turkey out of the Great War. One of the British ships was named *Agamemnon*, after the commander of the Greeks during the Trojan War; like that ill-fated king, Churchill too underestimated the resilience of an Asian power. His aims are well known – to take control of the waterway, secure passage for his warships, and occupy Istanbul. Two and a half thousand years ago, Herodotus had written that conflict between Asia and Europe was all the result of bride-stealing; in place of Helen of Troy it would be Constantinople – 'Queen of Cities' – that Churchill would attempt to steal back for Europe. But in the assault on the strait the French and British navies couldn't penetrate the minefields; three battleships were sunk and three more were crippled. The operation descended into the slaughter of the Gallipoli campaign: British imperial troops attempted a land invasion of the European side of the strait against a bigger army who held the high ground.

On 18 March 2022, the anniversary of that Turkish sea victory over the Allies in the First World War, the longest suspension bridge in the world was opened across the Çanakkale Strait between Asia and Europe. From one perspective this immense bridge, celebrated with all the pomp and splendour of a military parade, has at last reconciled the two continents, soaring over waves last bridged 2,400 years ago by Xerxes.

It is a bridge freighted with history. During the pandemic lockdowns, I watched on the internet as its tower rose and rose, to top out at more than 300 metres, blood-red against the dull khaki of its European and Asian shores. Numerologists have pointed out that the height of its towers, 318 metres, recalls the 3/18 (18 March) of Churchill's failed assault on the Strait; its length, 2,023 metres, recalls the centenary of the founding of the Turkish Republic. Workers, funders and engineers from Türkiye, Korea, Japan, Australia, UK, Netherlands, Denmark and India were involved in its construction, making it possible for the old Roman Via Egnatia to reach the heart of Asia.

When I visited the ruins of Troy in the mid 2000s, at the outset of a year-long traverse of Asia, work on Erdoğan's new bridge over the Hellespont had not yet begun. To cross back over to Europe and visit the Gallipoli peninsula I was obliged to take a boat. The jetty for the ferries of the strait lay alongside the stretch of shoreline where it's believed Xerxes anchored his pontoons. I watched the water run from the Black Sea to the Mediterranean, tight as a muscle, silvered as armour – water that now runs beneath a ruby-red bridge. Scintillations on the surface flashed like particles in a cloud chamber. Geologists say that veins of red granite and porphyry run uninterrupted between Europe and Asia underneath the Hellespont, and I thought of the indivisibility of those continents, and the wars that have been fought for the right to control these lands and this water – wars that may come again.

It's what we humans do, isn't it? Build connections even as we foster divisions. From the bulkhead of the little ferryboat, I threw a Turkish penny into the water and made a wish: for more wisdom in the lines we draw around ourselves, and in where we choose to build our bridges.

Chapter Thirteen

Bridge of Conquest

Kabul, Pakistan
Old Attock Bridge: steel truss, two-level (1883), 420m
Border: Punjab–Khyber Pakhtunkhwa

On the other side of the range were tribes and places, of which we had never heard the names . . . All we could learn was, that beyond the hills was something wild, strange and new, which we might hope one day to explore.

MOUNTSTUART ELPHINSTONE, *CAUBUL*

Peshawar's Heritage Hotel was furnished like a diorama from a museum of the British Raj, but had fallen on hard times. The beds were chipped and sagging, the chintz had been eaten by ants, and stuffed animals had been nailed to the walls of the staircase. Men were stationed at the door of the hotel, guns slung at their hips. Peshawar's men, I realised, wear weapons the way others wear wristwatches.

I had a romantic notion to reach the Khyber Pass, Pakistan's border with Afghanistan. Standing on the ramparts of the Hindu Kush, Afghanistan and Central Asia behind me, I wanted to look down on the Kabul and Indus valleys below, all the magnificence and immensity of the East and South Asia before me. The trouble was that E. and I had to obtain a permit to get there.

The Pakistani army were in the middle of one of their spasmodic purges of 'insurgents', and a Khyber permit, we were told, would be 'delayed'.

'Delayed for how long?' I asked the hotelier who was acting as our fixer.

He had watery eyes and carefully groomed moustaches. A wig balanced on his head like a chimney brush. 'Next year?'

The British Raj had called this region west of the Indus 'North-West Frontier Province' in acknowledgement of the fact that two bombastic wars had failed to conquer the Afghans beyond it. Early in the twenty-first century intense lobbying by local politicians secured its name-change to 'Khyber Pakhtunkhwa', a tongue-twisting drum-roll of syllables that Anglophone newsreaders have struggled with ever since. It is a tight mesh of valleys occupied by hostile tribesmen who now earn the respect of the Pakistani military as they once did British imperial officers. These Pathans, or 'Pakhtuns', have a long history as thorns in the sides of empire-builders: Herodotus wrote of 'Pactyans' causing trouble to imperial powers as early as 1000 BCE. Winston Churchill, stationed in the Hindu Kush as a young second lieutenant, called it a land of 'savage brilliancy'. It was in Churchill's interests to exaggerate the threats he faced, and in writing of the tribesmen he said, 'the ferocity of the Zulu are [*sic*] added to the craft of the Redskin and the marksmanship of the Boer'. The Americans who fought their way through these valleys on the Afghan side of the border were told to pack Churchill's memoir as required reading.

In the Peshawari bazaar there were placards advertising work and study visas in the UK and Ireland. Gangs of women in burkas fingered tables of nylon underwear. There were dusty gutters along the margins of each street that looked as if they knew the nocturnal attention of rats. Postcards of Kabul, Mazar-i-Sharif and Herat were piled in racks; remnants of the 1970s tourist industry when the Hippy Trail still poured down to Peshawar from Afghanistan, India-bound. The women in the postcard photographs had been captured by the camera on the eve of the Soviet invasion: they were bare-headed and confident, their eyes haloed with rainbows of eyeshadow. I tried to imagine what had become of them all after decades of Russian then American war, and now Taliban rule. As E. and I pushed through the crowds of beards and downcast eyes of modern Peshawar, hers was the only bare female face.

The Pathans were gracious and hospitable hosts, hungry for news of the

West; at each stall we visited we were asked to sit and take tea, to talk about where we had come from, what we thought of Pakistan, of Afghanistan and, of course, America. But behind the warm camaraderie there was a brittle tension. When a car backfired, the chatter of the bazaar would fall silent. 'Car bomb!' yelped one of our hosts, before giving a shrill laugh. The city was at that time averaging one car bomb or assassination a day.

We visited Peshawar on the sixtieth anniversary of India's partition. There was a grubby, mineral feel to the back streets of the town, as if the slums of Victorian London had sunk into the rocks of the Hindu Kush. Fear throbbed just at the threshold of the senses, it trembled in the streets and ricocheted between the abandoned palaces of Hindu merchants, now derelict. Streets were divided by category of merchandise: some for the sale of spices, others purely medicines, others jewellery and yet others butchered goats. It seemed as if all the riches of the Indies and of Central Asia could be had for sale in the Peshawar bazaar, you just had to know where to look.

One of the shadier streets sold guns and explosives, and we found American military rations for sale alongside state-of-the-art night-vision goggles. There were plenty of other looted supplies; it seemed that the Pathans were still getting the better of imperial invaders.

Between the spans of Attock Bridge, just east of Peshawar, the waters of the hot brown Kabul River mingle with the methylene-blue Indus. The two rivers flow in parallel for hundreds of metres as if afraid to lose their separate identities – the Attock Bridge leapt over the waters in uneasy coexistence.

The Kabul was rust-coloured, ferrous; it rises from the baked earth of central Asia and runs through the outskirts of Kabul city and Jalalabad before reaching Pakistan. The Indus, by contrast, seemed a fusion of water and sky; in its tinselled blue light it held a memory of the high country: Karakoram and the Tibetan plateau. From Attock the doubled Indus flows south across the plains of Pakistan, and was itself a border in antiquity

– pious Hindus believed that to cross to its western bank was to betray one's religion and risk outcast status.

The bridge I crossed over the twinned river was a road bridge of concrete spans, built in the last few decades, but just downstream I went to see the original, a simple frame of steel trusses, still in use for trains though built by the British Raj on two levels: road beneath, rail above. Its red paint reminded me of the Forth Rail Bridge of a similar era. A train whistle like a scream made me flinch.

In Peshawar I had been reading *Baburnama*, the perky and upbeat journal of a Central Asian warlord who, in the early 1500s, had conquered Kabul and then used it as a springboard for his conquest of India. Babur founded the Mogul Empire – a far more durable colonial project than the British one – and was the great-grandfather of the man who built the Taj Mahal. His grandson, Akbar, had a magnificent city built by the River Yamuna on the Ganges plain, Fatehpur Sikri, in which a monumental arch almost two hundred feet high was inscribed with one of the sayings of Jesus, as repeated in the Qur'an: 'The world is a bridge, pass over it, but build no houses upon it.'

But for all the piety of Akbar, his grandfather's journal has a rare candour for that of a leader or statesman, and implies a loose and sybaritic interpretation of Islam. It's full of descriptions of the lands he coveted and conquered, their worth in taxes, the skills of their men as soldiers, and appalling tortures meted out to men who displeased him. But it's also full of practical jokes, and accounts of booze-fuelled orgies ('Nasir Dost was so drunk that the men could not get him on his horse . . .').

Babur built a pontoon bridge over the Ganges to facilitate his conquest of Bengal, and explains how he managed to get an army of tens of thousands of men across the Kabul and Indus rivers – prefiguring the Attock crossing by more than three hundred years. But as I read his book in the heat of Peshawar, it was his account of the local region that appealed to me. He describes the land around the confluence of the Kabul with the Indus as a jungle teeming with rhinoceros, and recounts how he killed three of the animals in one afternoon. 'I have often wondered how a rhino and an elephant would behave if brought face to face,' he adds. Around me was only dust and heat haze, little in the way of greenery, and I couldn't help wondering how much of the despoliation of the land around us occurred under Babur's family rule, how much under their successors, the British, and how much under the jurisdiction of the free state of Pakistan – which means 'Country of the Pure'.

The air was filled not with jungle smells, but with particles of dust as fine as flour. Around the banks of the Kabul there were sprawling shanties of earthen walls. Sounds were of honking horns; the bridge itself has become an extension of the Grand Trunk Road that stretches fifteen hundred miles to Calcutta on the Bay of Bengal – a road that existed as one of the world's great highways before even Alexander the Great tramped down it. Kipling wrote of it: 'truly the Grand Trunk Road is a wonderful spectacle . . . such a river of life as nowhere exists in the world'. The first European maps of India, drawn in the early seventeenth century, mark it as 'The Longe Walke', giving its avenue of shade trees more prominence than the Himalayas themselves.

With Attock Bridge the Raj extended the Great Trunk Road from Bengal past the Indus and into the frontier lands of the Hindu Kush – the better to move goods and people around the subcontinent. Mohandas Gandhi, whose formative years were in British-ruled South Africa, began his career with the belief that these networks of empire could prove of benefit to the people of India. Over the course of his life he moved to the opposite position: that they were a scandalous imposition, intended to benefit only the colonisers.

What Gandhi had to say of British-built railways was also true of their bridges:

> It must be manifest to you that, but for the railways, the English could not have such a hold on India as they have. The railways, too,

> have spread the bubonic plague. Without them, the masses could not move from place to place. They are the carriers of plague germs. Formerly we had natural segregation. Railways have also increased the frequency of famines because, owing to facility of means of locomotion, people sell out their grain and it is sent to the dearest markets. People become careless and so the pressure of famine increases. Railways accentuate the evil nature of man: bad men fulfil their evil designs with greater rapidity. The holy places of India have become unholy. Formerly, people went to these places with very great difficulty. Generally, therefore, only the real devotees visited such places. Nowadays rogues visit them in order to practise their roguery.

For Gandhi, the British mania for the building of railway bridges had its origin in an impulse not towards the development of the country and the benefit of its people, but towards resource extraction, military efficiency, divide-and-rule. The ease of moving grain had the side effect of facilitating the spread of disease.

Dust joined with heat haze in a grimy fog. Scooters and minivans dodged around potholes, beeping and yelling and dropping their cargo. Cyclists veered onto the sidings for safety whenever other vehicles came too close. Brightly painted trucks, caparisoned like elephants and jingling with chains, grunted and kicked up dust. The trucks were adorned like the wagons of a travelling fairground, with gaudy colours and ornately carved wooden doors. Panels on their sides glittered with Bollywood actresses, images of paradise, Himalayan peaks and kung fu heroes. On the front of each was a galaxy of LED lights, and on the tailgates the incongruously polite request: 'Horn Please'. On Pakistani roads, a functioning horn seemed to be as important as a steering wheel or an accelerator. A taxi driver in Peshawar had pulled over and asked E. and me to step out. 'I'm sorry,' he said, 'but my horn is not working. You must find another car.'

Between the trucks, little taxis called 'tuk-tuks' for their smoke-spewing two-stroke engines clogged the highway. In the strict pecking order of Pakistani roads, they were on a par with us riding our motorcycle.

The Afghan trucks driving down from the Hindu Kush rolled on oblivious to our presence, and I watched and learned from the way the tuk-tuk drivers swerved into the gutters at the side of road, dodging fruit stalls and cyclists, to let the larger trucks pass by. For Kipling, in his pride at the jewel in Britain's crown of empire, the Grand Trunk Road was a river of life. But for us on our motorbike, the dust, the absence of greenery, the parched hazy air, the wrecks littering the sides of the road made it seem more like a gutter of death.

Chapter Fourteen

Bridges of Partition

Indus, India
Srinagar–Leh highway bridges: steel / timber / concrete beam (1962–4)
Border: India–Pakistan

There can be no question of coercing any large areas in which one community has a majority to live against their will under a Government in which another community has a majority – and the only alternative to coercion is partition.

LORD MOUNTBATTEN

The Wagah border just east of Lahore was the only legal crossing between Pakistan and India. We had arrived by driving down the Great Trunk Road, and were hoping to get north to the mountains on the Indian side of the border before the heat became intolerable. Punjabi soldiers from both sides of the divide patrolled up and down a line marked on the road, their bodies erect as raised swords. Each of them was basketball-player tall, though like cockerels they augmented their height even further with fan-like wattles that jutted from the peaks of their hats. Every evening, during a ceremony to mark the lowering of their flags, they snorted, stamped and bellowed at one another before an audience that could reach thousands. We watched from the Indian side.

The heat was feverish, so humid it was submarine; silky and mingling with sweat into one liquid medium. On the Pakistani side a sea of cream and beige shalwar kameez, segregated by sex, waved green star-and-crescent flags. On the Indian side there was beer for sale, and women dotted through the crowd in saris as colourful as rhododendron flowers. Above them was a sign: 'Welcome to the World's Largest Democracy!' As the soldiers marched the crowds roared 'Pakistan Zindabad!' or 'Jai Hind!' until the air between the trees trembled.

It's thought that a million people were killed when the Punjab was divided and the border at Wagah was created. Fifteen million refugees were displaced. Behind the spectator stands for the border ceremony runs the Wagah railway, where in August 1947 eastbound trains loaded with Hindus and westbound trains loaded with Muslims were boarded by murderous gangs who slaughtered everyone on board. I looked at the trees that grew thick around the stands, and thought of all the blood that has nourished their roots. The patriotic yelling of the Wagah ceremonies feels like a substitute, a compensation even, for the violence that erupted here so recently.

A few miles on from the Wagah border E. and I reached the River Beas, and crossed it over a modern concrete bridge, the waters corralled by cement-grey levees. 'Punjab' means 'five rivers', and this was one of them; though the structure of the bridge was unassuming, the river it crossed was once the eastern frontier of Alexander the Great's empire. It was on the west bank of the Beas that Alexander's generals refused to march any further and he was obliged to concede defeat, not by one of his adversaries, but by his own army.

His biographers said he had giant houses constructed here, complete with outsize beds and chairs, so that all who came afterwards would marvel at the stature of the Greeks. He also had a monument raised, on it the legend 'Alexander Stopped Here'. Now there was only the concrete highway, a few scrubby fields. It was a late-August day, hot on the plain, and I slipped down to cool my face in the water.

A couple of weeks later we were in the high Himalayas, along the liminal frontiers between India, Pakistan and China. We had driven the motorcycle from the banks of the Ganges north across the mountains, intending to reach Ladakh and the valley of the Indus. There was a road there that I wanted to drive, from Ladakh to Kashmir, which had been constructed in the 1960s. The partition of India had amputated Ladakh from its natural trade routes into the Karakoram, which now lay in Pakistan. India needed new access roads and bridges, but these had to be built across high, arid desert terrain which is buried in snow for much of the year. The Kashmir–Ladakh road is an engineering achievement, and also an anomaly; a series of bridges to mark the failure of diplomacy, made necessary by the imposition of one of the most tense, contested and militarised borders in the world.

The horizon was corrugated in jagged silhouettes, framing a depthless and luminous vault of sky. In the lower valleys there had been mists, and rainbows like gasoline spills on water, but up beyond the treeline the light bearing down had an abrasive, crystalline quality; it scored lines and shadows into the landscape as if with a diamond blade. The mountains that rubbed shoulders there were simple in their immensity and, from a distance at least, majestic in their beauty. But up close they were nothing more than undulations in an airless desert plain. Heaps of grey and rust-red stone lay tumbled across the landscape. It looked extraterrestrial, like a Martian upland, or some arid lunar sea.

The makeshift tracks we rode on had been hammered out by the passage of Indian army trucks. Soldiers patrolled the nearby disputed border with Tibet, staring down their Chinese and Pakistani equivalents and driving with a fatalism that would be admirable were it not so dangerous for everyone else on the road. Car and truck wrecks littered the valley floor, and I watched bulldozers work the mountain to keep the road open against landslips. To a motorcyclist, the military truck drivers were the greatest threat; high on amphetamines they drove in pitiless and murderous convoys. E. began to gather rocks in her pockets, to throw at them as they approached in a fruitless attempt to get them to swerve out of our way. One truck came so close that a prong jutting from its tailgate ripped open our baggage.

Somewhere ahead lay a caravanserai marked by a scribbled cross on the map, a transient summer camp 4,500 metres into the sky. At this altitude the colour chosen by the cartographers is white: white as an ice field, as a polar plateau. But we were closer to the tropics than the poles, and there was no ice on the road. My breath came in gasps that seemed to fill but never satisfy my lungs. There was a trembling ache in my arms from holding the handlebars, and I was fighting back the nausea of a headache that drilled at my temples. It wasn't altitude sickness yet, but it would be if I didn't descend soon. E. sat pillion behind me, and I realised she'd been silent for the past hour. Her weight was negligible among all the steel and luggage that ballasted us, and a sudden panic stabbed through my mind – perhaps she had fallen off. I took my left hand from the handlebar and reached back to touch her knee. She squeezed my hand in return. I kept hallucinating her absence; the lack of oxygen was playing tricks with my mind.

The bike with its fuel and luggage, together with our weight, came to nearly half a tonne. All day I'd steered that weight through riverbeds, across landslides, up boulder-strewn sidings and over a pass of 5,400 metres. Fatigue trickled into my limbs like molten lead, scorched at the surface of my eyes. I hadn't thought we would make it as far as we did, but at each gruelling stage reached, the thought of turning back was always worse than the idea of going on.

At first we swerved through narrow labyrinths of purple and maroon, their strata twisted by the tectonic forces that had buckled this whole range. The valleys held their breath, as if waiting for a landslip. We hoped to make the caravanserai before nightfall. Truck wrecks occasionally blocked the road, and at the base of one of the steepest climbs we met a party of Sikh traders whose disembowelled vehicle lay in pieces across the road. A makeshift path had been carved out over the roadside rubble to avoid it. They smiled and offered tea; for two weeks they had been waiting on the arrival of replacement parts.

We approached a series of gorges crossed by simple beam bridges resting on dull concrete supports, and as the track dropped into them the luminosity of the sky shifted from quartz to a dull ruby. We were accelerating

downhill, negotiating a labyrinth of rock passageways and swift-running riverbeds, racing the advancing flood of night as the air thickened a little and my headache began to recede. The grandeur of the Himalayas around us fell away as the mountains – only an hour ago so vast – faded to voids within a canopy of stars. We were edging forward, the headlight feeling its way over the rubble like a blind hand, when ahead I saw the glow of two tents.

An Indian boy at a stove inside was preparing noodle stew. We were high above the clouds but even here there was an obvious caste divide. He looked barely fifteen, round-cheeked and dark, with eyebrows like chevrons over restless, inquisitive eyes. He darted around the space, cooking and cleaning, while pale and angular Ladakhis sat and talked, wrapped in their blankets. Theirs is a dialect of Tibetan, a language more closely related to Burman and Chinese than to Sanskrit, rumbling and percussive like a riverbed in flood. The Hindi the boy spoke was lighter, airier: it sang and splashed along the surface of his voice. He grinned and gave me a nod of fellowship. 'My name is Arjun,' he said in Hindi, touching his right hand to his heart, then he added in English, 'I am not from this place.' His language skills put me to shame; later I heard him speak Ladakhi, and even some Chinese.

His noodles were delicious, but his butter tea was rancid. We gulped it down anyway as fortification against the cold. When we handed the bowls back he nodded towards a small cubicle, separated from the open tent by some grimy curtains. A stack of blankets was piled against one canvas wall, and he shook two of them out. Plumes of dust obscured the light from the hurricane lamp. We wound scarves around our faces as masks and, still wearing all our motorbike gear, wrapped ourselves in the blankets and lay down.

It was dark, and the sounds were the hiss of the stove, the murmur of Ladakhi, a gentle wind flapping the canvas. No one beyond the walls of this tent knew where in the fastness of the Himalayas we had buried ourselves for the night. The grandeur of the mountains slept in darkness, and above them the great planetarium of the night sky turned. Beyond the tent the sky was as silent as those seas on the moon, Mare Serenitatis, Mare Tranquillitatis. After an hour or so of sneezing, coughing and gasping, sleep came like a smothering of wool.

There is a brief season in summer when it's possible to work on the road to Ladakh, also the only part of the year that it's passable for traffic. The next day we encountered road gangs of Nepali and Bihari workers, tar-stained and in rags, filling in the road's potholes and mending the decks of its bridges. For trucks this means lengthy waits, while the earth-moving equipment and tarring machines do their work. For the thousandth time, E. and I blessed the motorcycle – it was enough for us to leave the road and drive down through a dry riverbed or over a rubble siding in order to get past the gangs. E. grew used to leaping off the pillion and walking ahead to find the best route over the stony ground.

The Indus is nourished by icemelt from the highest peaks, smuggling rock dust from the summits and carrying it thousands of miles to the

Indian Ocean. When it reaches the sea that dust will, in a sense, be going home – the bones of the Himalayas, all the way to the summit ridge of Everest itself, were once part of an ocean floor. Through the force of a continental collision they have been dragged up from the abyss in a tectonic crunch that sees India eliding with Eurasia at geological warp speed – one metre every twenty years. These are young mountains, geologically speaking – only around forty million years old – and they are still rising by about a centimetre every couple of years. The squeezing pressure of the subcontinent has crumpled the Tibetan plateau so much that its crust is now double the thickness of the rest of the planet, and distorts the earth's magnetic field.

Suddenly, from behind a wall of shattered stones, the River Indus came into sight. It flowed from east to west down a narrow flood plain lined with trees and rough grazing – that modest smear of green was a delight to the eyes after the grey rock of the high passes. Its water was milky with silt and dust, and ran swiftly towards the town of Leh. The bridges were of simple military concrete, with beams and chipped uprights; occasionally they were of steel trusses, with a deck of wooden planks. The latter I had to drive over with care – there were many nails sticking up out of the timber, and some of the joists were rotten. Simple block houses began to appear, their only extravagance streamers of prayer flags that wagged like tongues praying for the salvation of all beings. Ladakh's architecture is distinctly Tibetan in style; it has historical ties far closer to Tibet than to India. Conquered by the Sikhs in the mid 1800s, it was enveloped by India only because the English then promptly conquered the Sikhs.

Children ran through the steep back streets of Leh's old town, swooping and chirruping like swallows. The houses were earthen and cracked with drought. Stacks of pancake dung for winter fuel were piled along each wall – it was not clear where the dung ended and the bricks of the houses began. From the summit of the town's fortress there was a brisk clarity to the air; the wind whistled through ribbons of prayer flags. Though Leh is buried deep in the buckle of Asia's greatest belt of mountains, the feeling was of trade routes spreading as far as Bokhara and Delhi, Beijing and Tehran. Tibetan monks walked the streets, beside devout Muslims in white robes and skullcaps, and bindi-daubed Hindus from Kerala and Tamil Nadu. My mind felt high, clear and free, bouncing around this roof of the world. To the north was the Nubra Valley, where the highest road in the world passes nearly six kilometres above sea level. To the south lay the fairy-tale peaks of Zanskar and Stok; improbable serrations of lilac and silver.

It was a Sunday, and there was a church service going on in the Moravian Mission – a sect that began far away in the valley of the Danube. The hymns of the Sunday service rose up into a clear space between the mountains.

Above the village of Lamayuru a monastery irrupted from the mountainside like a molar on a bloodless gum. I met a monk in burgundy robes who had been born in the village but had travelled far beyond these valleys. His hair was cropped short, and his eyes were like amber seaglass, stippled and opaque.

'My parents were unhappy when I decided to become a monk,' he said. 'They hoped that I would marry and give them grandchildren. But now they realise that I had no choice – no other way of life was possible for me.' He poured tea. We had been sitting for some time. He had little opportunity to speak English, so he talked, and I listened.

'I should never seek to convert others to Buddhism. An individual must ask me three times, and then perhaps he is ready. To speak of the teachings of Buddha before a pupil is ready is like pouring water onto an upturned bucket. It is wasteful. When someone is ready they are like a bucket the correct way up. The Dalai Lama himself tells you Europeans that you should stick to your own traditions of compassion, your Christian ones.'

'I heard the Dalai Lama teach once,' I told him. 'He said, "Your Western traditions are as valid as ours."'

'So you have seen His Holiness,' he whispered, eyes widening.

He hinted that he had travelled beyond India, and I asked him where.

'For seven years I studied Buddhist scriptures in Taiwan. The Chinese language is not so difficult as they say.' He shrugged, as if moving to Taiwan and learning ancient Chinese scriptures were routine endeavours, easily accomplished. 'Now I will spend seven more years learning the Tibetan scriptures. Then perhaps I will become a healthcare worker. The Tibetans who come to India, escaping the Chinese, are very sick when they arrive. Learning how to treat sickness is good, a most noble work.'

Behind him, to the south, there were valleys that climbed steadily towards the ice caps of Zanskar. 'It's beautiful here,' I told him. 'But hard in winter, I imagine.'

'Hard, yes, hard and poor. Why do you Europeans come here? You come so far from your homes and families, to walk up our mountains. What for? Why do you not stay in your own countries, climb your own mountains, further your knowledge of the places where you belong?'

A couple of hundred kilometres a day was the most I could ever contemplate driving on those roads, heaped as they were with ungraded rubble so rough it appeared to have been bulldozed into place directly after being dynamited out of the earth. Leh to Kargil was a trip slightly longer than was comfortable, 220 kilometres, and as E. and I drove the final few miles, my limbs began to tremble with exhaustion. The motorbike went over a hump and as it came down on its suspension, a sharp rock that I hadn't been alert enough to avoid breached the engine's sump. The bike began to haemorrhage oil, inking a black line along the track behind us. I flipped the gearbox into neutral and coasted down into Kargil, arriving within minutes, by a stroke of luck, at the only place within a hundred kilometres with a mechanic capable of fixing the leak.

The Indus had branched off a few kilometres back, towards the Skardu Valley and Pakistan. The river that divides Kargil is a tributary of it, called the Suru. The motorbike rolled silently to a halt a hundred yards or so from a mechanic's workshop. E. and I pushed it the last few steps.

* * *

Kargil might lie on the front line of Indo-Pakistani conflict, but it also lies on one of the world's great cultural watersheds. To the east it stands guard over a dividing line, a border that is felt rather than seen, where the tide of Islam, having flowed uninterrupted all the way from west Africa, washes up against the ramparts of the Himalayan range and begins to turn back, giving way to Buddhism, Confucianism, Taoism. Beyond it the Muslims are the visitors and the Buddhists feel at home. At Kargil, the road veers close to the ceasefire line with Pakistan. In 1999 the tension here had boiled over into a summer war – a war brought about in many ways by the presence of the road. The world watched while fighter jets flew through the valleys of the Indus, accelerating tensions between two nuclear-armed states. The new Kashmir–Ladakh highway had been shelled from the Pakistani side, and was still pockmarked in places. As we drove west we passed roadside signs declaring 'WARNING! You are under enemy surveillance!' There were more army trucks, soldiers in bunkers, and Sikh, Hindu and Muslim memorials to the dead. Camouflaged mortars aimed from the roadside into the mountains, at known Pakistani positions. Spindled steel bridges had been spun across the Indus tributaries. I watched the water draining off towards the Pakistani side, thinking of how millennia of natural trade routes and cultural connections had been severed by this conflict.

The sky was opaque, brown at dusk; the mountains' edges became a torn paper silhouette. Then night came like a coffin lid pulled over the sky. There was no street lighting after sunset – perhaps Kargil's people feel safer in the dark. That they were under surveillance by Pakistani artillery positions gave the streets a hostile, hysterical edge. It felt as if there was a war drum beating just below the threshold of hearing, prickling the hairs on my neck, trembling the air with a timpani of fear. We took cover in an otherwise empty hotel, and ate a solitary curry by candlelight. The room had pillbox windows high on one wall. Shrill screams and the barking of dogs broke the silence of the night. In that room I dreamt of artillery blasts, and saw bodies tumbling down the mountainsides into the waters of the Indus.

In the morning we stopped at an old bridge that crossed the Drass River, which runs close alongside the militarised frontier with Pakistan. The road beyond it had been unused since the establishment of the line of control in 1949. The bridge was barricaded with barbed wire and there was no guard, though plenty of signs informed me that I was under surveillance from both Indian and Pakistani armies. It was of course forbidden to cross. A bridge that led nowhere, a useless bridge, a monument to partition.

West of Kargil the tyres of the motorbike cracked ice on the puddles, though it was late summer. From the Zoji Pass the road descended towards a sullen collection of low stone huts and over a forlorn, poorly maintained bridge rattling with loose wooden slats. The rocks in the riverbed beneath me glistened pale as bones. An Indian government sign informed us: 'Drass – Second Coldest Inhabited Place on Earth'.

In winter the temperature here can reach 60° C below zero. The power of altitude was such that, though we drove through tropical latitudes, it was as if we had been transported high into the Arctic Circle. There were anaemic fields, wan-looking trees, and pale-skinned, green-eyed people carrying baskets of fodder. The roads in and out of the town can be closed for eight months of the year with snow.

The silver gleam of the river was a keel along the steep valley floor, ballast to the ship of the valley. Outside a hut by the road to Srinagar, between stalls of bearded goats, I met a flame-haired Glaswegian who had first come to India as an air hostess. Gillian spoke with a burred mid-Atlantic drawl and her eyes were of the sea; blue, restless and glittering in the sunlight. Somewhere along the way she had married, divorced, trained as a photographer, washed up on the beaches of California, and found a job as a college tutor. For ten years she had divided her time between Kashmir, San Francisco and Strathclyde.

'Why Kashmir?' I asked.

'It chose me,' she said.

Ten years previously she had been in Delhi, on holiday with her sister, when the heat, poverty and pollution of that city had become too much for them both. They booked bus tickets for Srinagar. 'In those days no one came up here,' she told me, taking in the wooded slopes and the five-thousand-metre peaks around us with one wave of her arm. 'The army were gunning down people in the streets, so we had the place to ourselves.'

What had been a quick visit for her sister became an obsession for Gillian, and within a year she was back. She always returned to the same village; the locals grew to expect her. As a foreigner she was exempt from the usual taboos and expectations that bear down on Kashmiri women. Now she owned a horse, half a guest house, and a houseboat on Dal Lake in Srinagar.

A herd of Kashmir goats, bells around their necks, ambled past in a gentle tintinnabulation. They were followed by a flock of skipping children in knee socks and striped ties; Enid Blyton uniforms in the high

Himalayas. Gillian offered them a smile as sunny as the dimple on an apricot, and said a few words in dialect – they giggled in reply, covering their mouths and hurrying down towards the village mosque. 'I tried to build a guest house here,' she said, and pointed at the shell of a structure further up the valley, its doors and windows empty. 'I thought it would be good for the economy. But every year when I leave for the winter, the building work stops.' She rubbed the soles of her boots in the dust. 'People here have no head for money, no idea about capitalism. And why should they?'

Tea was brought from one of the low stone houses, clattering on a Johnnie Walker tray. I nodded my head and took one of the little glasses of brown liquid piled with sugar cubes. Its bearer, Aziz, beamed at Gillian and nodded silently towards E. and me. A convoy of khaki jeeps came around the bend, blasting their horns to clear the goats from the road as their engines screamed. The herdsman panicked, flailing his arms and yelling at the animals to move to one side. In a moment the jeeps were gone, and the herder ran up and down trying to gather his goats.

'Army,' said Aziz, and scowled.

'The people here hate them,' added Gillian, 'and many would dearly love to join with Pakistan. But without tourists and Hindu pilgrims from India, I'm not sure they'd survive.'

Gillian had been drawn to Kashmir and Ladakh by its relative emptiness, an emptiness that was a consequence of the conflict. We too had only been able to cross those mountains by motorbike thanks to a series of military bridges thrown over some of the most challenging terrain on earth.

From Ladakh we dropped into the Vale of Kashmir, and after a few days negotiating checkpoints staffed by the Indian soldiers of Srinagar, drove back to the plain of the Beas River. We had come full circle, back to the Punjab where India and Pakistan had been divided, close by that line where Alexander had his frontier.

Indian military expenditure now approaches USD $80 billion per year; the education budget for the second most populous nation on earth is just a quarter of this figure, and on health, the country spends just a fifth. Our journey along the partition line, among Kashmiri and Ladakhi people who've been cut off from their natural lines of connection, made me wonder what kind of world it would be if those figures were to be reversed.

Chapter Fifteen

Bridge of Peace

Xiaohe, China
AnJi Bridge: stone single-arch (605), 50m
Border: dukedoms of Ch'i–Han

There is nothing in the world more soft and weak than water, and yet for attacking things that are firm and strong there is nothing that can take precedence of it.

TAO TE CHING

E. and I spent ten weeks journeying around China, from the borderlands of Kazakhstan to those of Korea, from the Yellow River in the north to the Red River in the south. In the Turfan Depression – the lowest dry point on earth after the shoreline of the Dead Sea – we followed *karez* channels hollowed out of the desert and roofed over with stone, so that snowmelt could run across burning sands with minimal loss to evaporation. In the back streets of Turfan we met an Uighur imam, and E. conversed with him in Arabic about the reach of that language, Morocco to Mongolia. Along the frontier with Inner Mongolia we walked on one of the most impressive border symbols to be constructed anywhere in the world: the Great Wall. Everywhere I had the sense of the motor of Chinese history turning, felt but unseen, a sense heightened by the knowledge that China was defending its borders when Europe was still populated by roving tribes of woodfolk. China astonished me with its diversity, with the antiquity of its human settlement, with the majesty of its deserts, forests and highlands, and even its densely squalid coasts. Everywhere we went there were stupendous bridges: in Hong Kong and in Shanghai, in Wuhan and in the steadily rising dam of the Yangtze.

E. abhorred China. Fluent in five languages, she had always assumed that a phrasebook, her linguistic skill and a degree of human curiosity would see her through – and was proved wrong, wrong and wrong again. China puzzled her like an impossible fraction, $\sqrt{-1}$. I speak only English and Italian and didn't attempt to communicate in Mandarin beyond pointing at phrases in a book. But I fell under the country's spell nevertheless.

Historians of language tell us that along a line between Bengal and Donegal most Indo-European languages are related, and though there are exceptions to that rule, there has been continuous trade and traffic between those places for many centuries. China by contrast has for long periods of time closed itself off from engagement with the north, south and west; it has deliberately turned away from connection. It was a delight to discover the extent to which its people had developed an utterly different approach to language, literature, government, religion than all of those I was accustomed to. For me it felt like a new-old world, gifted with what seemed a parallel approach to being human.

China's Szechuan province has been building suspension bridges of bamboo for many centuries. Two of the Yangtze's most ample tributaries meet at Leshan, in a torrent of whirlpools and shingle beds that churn at the feet of one of Asia's largest rock-cut Buddhas. The location has long been considered celestially blessed, conducive to healing, and in the parks along the riverbank old folks were wrestling with municipal gym equipment and moving in synchrony with the slow dance of t'ai chi. The water level was high, and I watched as a succession of swimmers donned inflatable arm bands, then threw themselves into the river, swept up by healing waters. The current had a violent undertow that could suck you down and spit you out further downstream. The swimmers were trying to reach the Buddha's toes at the farther shore, and only a few made it.

Those caught by the current were hauled back in by safety lines tied around their waists. Without a bridge, their efforts to push across the current was an integral part of their pilgrimage – both mortification and penance.

One Chinese myth imagines a bridge known as Nai-ho-k'iao standing between life and death: in this tradition the dead will be obliged to drink a potion to induce forgetting of their earthly lives, then begin their crossing of the bridge. But they cannot help but fall into the torrent below, which will divide them into streams of potential reincarnation – men, animals, birds, fishes or insects. No Charon or ferryman transports souls across the divide between worlds. Just like pilgrims struggling through the waters at Leshan, only individual effort can secure passage across the divide into a blessed afterlife.

By around 300 BCE China had already developed suspension bridges of chains, and zipline bridges of rope for which wayfarers would carry their own handpieces. The country at that time was still divided between dukedoms. They seem small when their borderlines are laid over a modern map of China, but were vast countries in comparison with those in Europe at the time. Here are some of their names: Shu, Ch'in, Han, Yüeh, Ch'u, Sung, Teng, Lu, Wei, Yen. It was the Han who came to dominate, and by the 220s BCE, they had moved from their small heartland around Ch'ang-an (now Xian) to rule most of the Yellow River basin and some of the Yangtze – what is today still the heartland of China. By 250 CE the Han dynasty had fallen and the map fractured again; new borders appeared between the lands of Wei, Wu and Shu, then between northern Ch'i and Chou.

It took until 600 CE for another power strong enough to dominate all the others to arise: the Sui dynasty. Its rule was short-lived (it endured only 40 years), but by bringing the country under one rule for the first time in centuries, it proved transformative. The Sui conquered northern Vietnam, opened diplomatic relations with Japan, and began the construction of a vast network of irrigation and transportation canals between the Yellow River and the Yangtze. It was under their rule that one of the finest architectural monuments to be found anywhere in the world was constructed: the AnJi, or 'safe crossing' bridge.

It's a magnificent piece of engineering; a shallow-arched bridge of stone reaching 40 metres across the Xiao River in a single span, constructed at a time when many of the short-arched Roman bridges of Western Europe were beginning to fall into disrepair through neglect. No one knows how a Chinese medieval engineer was able to conjure this bridge in the imagination, or master the geometry required to build its height-to-span ratio of 1:5 – far beyond the ability of the Romans, and something that wouldn't be seen in Europe until the late Middle Ages (it has the same depth of arch as the Ponte Vecchio in Florence). Spandrels are those parts of a bridge that fill the space between the arch, deck and abutments (from the Latin *expandre* – to reach or stretch), and the AnJi bridge has hollow spandrels to allow floodwaters to pass through, giving it eyes that glare down at the water. The arch is composed of 28 rows of individually carved limestone blocks or 'voussoirs', each one bound to its neighbour with cast-iron staples. Studies of the rock have revealed they were quarried 30 kilometres away, and it is thought they were pushed to the bridge site during winter, when the Xiao River would have been frozen.

For a bridge so ancient, it looks today as modern as ever. The elision of borders brought by the Sui, and the subsequent peace, must have brought many advantages – the bridge not least among them. And it must hardly have been completed before that dynasty also fell – to be succeeded by the T'ang dynasty with its famous flowering of culture and art. T'ang officials weren't responsible for commissioning or building the AnJi bridge, but according to the scientist and sinologist Joseph Needham, they had it inscribed. The following was written in the eighth century:

> This stone bridge over the Xiao River is the result of the work of the Sui engineer Li Chun. Its construction is indeed unusual, and no one knows on what principle he made it. But let us observe his marvellous use of stone-work. Its convexity is so smooth, and the wedge-shaped stones fit together so perfectly . . . How lofty is the flying-arch! How large is the opening, yet without piers! . . . Precise indeed are the cross-bondings and joints between the stones, masonry blocks delicately interlocking like mill wheels, or like the walls of wells; a hundred forms in one. And besides the mortar in the crevices there are slender-waisted iron cramps to bind the stones together. The four small arches inserted, on either side two, break the anger

> of the roaring floods, and protect the bridge mightily. Such a masterwork could never have been achieved if this man had not applied his genius to the building of a work which would last for centuries to come.

With the 'safe crossing' bridge, the genius of the Sui has outlasted their dynasty by many centuries. Like the medieval bridges of Europe, it was considered a place of power and magic, and its masons had protective motifs carved into its rock. I walked from one bank to the other, from one world into another and back again, marvelling at the elegance of the stonework, the achievements of Chinese engineering, the durability of this crossing, the potential consequences of peace.

Chapter Sixteen

Bridge of Commerce

Singapore
Cavenagh Bridge: cable-stay (1869), 79m
Johor–Singapore: causeway (1924), 1,056m
Borders: commercial–civic centres, Singapore; Malaysia–Singapore

The fallout of the assault on our planet is impeding our efforts to eliminate poverty and imperiling food security. And it is making our work for peace even more difficult, as the disruptions drive instability, displacement and conflict.

ANTÓNIO GUTERRES, UN SECRETARY GENERAL

Sumatra and the equator lay across the narrow Malacca Strait, invisible through the dense haze and the traffic pollution. Where the litmus paper of the sky dipped into the ocean it turned a dull, smoky orange. On the banks of the Mekong, at the southern tip of China, I'd once seen the jungles of South East Asia begin; in Singapore, two thousand kilometres due south of that border, they reached a full stop. Centuries ago the Portuguese and then the Dutch had built trading entrepôts and docks further north on the peninsula, but an Englishman, Stamford Raffles, saw the strategic perfection of the island, and in 1819 negotiated with the Sultan of Johor to site a trading port at Singapore. By 1867, with England flexing colonial muscle across much of the world, it became a crown colony, having grown from a population of a few hundred to more than a hundred thousand.

Singapore is now one of the richest countries in Asia, a concrete jungle famously strict in its law enforcement (No chewing gum! No spitting in the street!) and a paradise of free-market capitalism. It glitters like a jukebox. The mountains of the eastern Himalayas as they curve towards the sea made overland trade between China and India all but impossible before the railways were built, so Singapore is the pivot on which, for two hundred years, trade and shipping between the two nations has swung – a nexus of global trade. Just as Venice, with its monopoly on shipping in the Adriatic, once made fortunes from its position as the bridge between European and Byzantine empires, Singapore, perfectly situated between India and China, now swells with wealth.

In 1941 the reporter Martha Gellhorn travelled through China for four months, reporting for *Collier's* magazine on the war between China and Japan, and the evolving civil conflict between the communists under Mao and the Chinese nationalists under Chiang Kai-shek. At the close of her journey she flew out to Singapore, arriving just a few months before the Japanese invaded: '. . . all the charming chaps in uniform fresh from Britain to defend that bastion of Empire. [Their] gaiety might have been on the feverish side. I think the Dutch consciously and the British unconsciously sensed that they were living in the last act before the fateful curtain.'

The Japanese army invaded the Malay peninsula in December 1941, and advanced quickly towards Singapore. A causeway between Johor on the peninsula and the island of Singapore had been completed in the 1920s; the British were obliged to blow it up. Japanese soldiers simply crossed on inflatable boats, then threw a bridge of girders down over the breach.

The defeat at Singapore led to the largest surrender of troops in the history of the British Empire – 130,000 men were captured and transported for slave labour, many to die on the infamous Burma Railway (an experience fictionalised in the movie *The Bridge on the River Kwai*). In the nearby sea battle for control of Malaya, the British navy suffered its biggest defeat of the war.

In that defeat it's possible to see the seeds of the rapid decolonisation that followed through the 1940s, 1950s and 1960s – India, Africa, the Middle East, Malaysia, and the change in status of Canada and Australia. I've heard it argued that the twenty-first-century independence movements in Scotland and Wales, and for the reunification of Ireland, are nothing more than the continuation of a retreat from Britain's empire – a process that arguably began with the Japanese victory in Singapore.

E. and I reached Singapore across the causeway, having our passports stamped at the frontier, and made for the southern end of the island. We had been on the road for more than a year, and our families back home were starting to wonder if we'd ever return to Europe. From Singapore I called my parents, and sent a picture postcard to my brother of a bridge that, like us, had travelled all the way from Scotland.

Its arrival in South East Asia was facilitated by the mid-nineteenth-century explosion of global commerce. The Cavenagh Bridge, named for a British governor of the city, was originally called the Edinburgh Bridge for the Duke of Edinburgh, and was commissioned to celebrate Singapore's colonial status as a crown colony. Oddly enough the 'Edinburgh' bridge was constructed in Glasgow before being shipped out to Singapore in parts. It was intended to connect the civic and commercial parts of Singapore. Following a century of colonisation, one Japanese occupation and 60 years of Singaporean independence, it does that still.

A cable-stay bridge in miniature, its supports are of lengths of cast iron locked together with rivets, rather than cables. I bounced across it, and over the Singapore river, thinking of a similar bridge I'd crossed seven thousand miles away over the Tweed. It was a kit bridge, Meccano-like and pleasing in its toytown simplicity. Its presence here in Singapore was testimony to the reach of an empire.

In the early 1960s Singapore was in union with Malaysia, but its Chinese majority began to feel threatened by new post-colonial laws that gave priority to Malays. Violence broke out, and in 1965 Singapore was expelled from the union. Its founder leader, the lawyer Lee Kuan Yew, was said at first to be distraught – he had hoped to strengthen the union. It's surprising now to recall that Singapore was forced to become an independent nation against the will of both its people and its politicians. As a tiny new country it was anxious and uncertain about independence, but there are few Singaporeans today who'd consider the creation of that border a mistake.

At the end of that year-long traverse of Asia, we found there at the continent's tip a museum of Asian civilisations. In it E. and I explored a distillation of the stories of the most powerful cultures between the Yellow Sea and the Black Sea. The museum had recently held a celebration of Cavenagh's Bridge and its symbolism – a bridge built at the inception of a truly global network of trade, forged on one continent to be assembled in the colonised territory of another. Its designers were the Glasgow engineering firm P. & W. MacLellan, and a descendant of the firm's founder, Gavin MacLellan, had travelled from Glasgow to Singapore to give a lecture. I flicked through his bio: MacLellan's career was one of global trade, exporting metalwork around the world, including parts of bridges and the tools to build them.

Many of the Chinese merchants who had settled in Singapore under the British were Cantonese, and among the temples they built I visited one to A-Ma, Cantonese goddess of the sea, protector from storms and extreme weather events. An effigy of the goddess stood guard at the temple's entrance, in silk and scarlet robes, a headdress of gold, her hands lost deep in imperial sleeves. She had a pacific expression on her face – her eyebrows were like 'o's with grave and acute accents: óò. Her presence here betrayed the kinship of Singapore with Hong Kong and Macau, and its allegiance to a borderless culture of seafarers and merchants, whose livelihoods mock the lines we draw over land, and whose networks of trade now extend to almost every part of the globe. Her gaze once looked out in the direction of waves but now, with the reclamations of earth of such a land-hungry state, the shore stood two kilometres away.

But the sea is coming back – without a drastic reduction in CO_2 emissions, sea levels are predicted to rise by one metre over the next eighty years, and four metres over the subsequent two hundred. Like many others of the world's most populous cities, Singapore will be submerged.

The city felt like a punctuation mark at the end of Asia. It was 2007, and CO_2 was at 384 ppm in the atmosphere. Worry was beginning to build within me about the carbon footprint of my own journeys, and I was starting to choose ways of mitigating and reducing that footprint. But as with almost everyone else on the planet, my choices were not drastic enough: in 2024 CO_2 reached 425 ppm, a level last seen more than four million years ago. Wildfires triggered by rising temperatures are a regular item in the news. Summer temperatures in Siberia approach 40° C – levels never previously recorded. Permafrost is melting, releasing methane, which though shorter-lived than CO_2 is a more powerful greenhouse gas. Soil holds more carbon than trees and plants put together, and it is increasingly obvious that we are losing soils too.

The planet has tried to help us out: about two-thirds of the CO_2 we have released as a species has been dissolved by the oceans (helping to balance the atmosphere, though it is slowly making the seas more acidic). Standing at the tip of Asia, watching the leviathan shipping vessels ease themselves around this pivot of international trade, it seemed as if the juggernaut of global networks of exploitation were unstoppable.

It is now inescapably obvious that we somehow have to get all that excess CO_2 back into the ground. If that ambition proves impossible to achieve, we may be obliged to explore drastic measures to cool the planet: perhaps to shower the upper atmosphere with dusts in order to dim the sun – geoengineering on a planetary scale. The climate scientist Andy Parker summed this up: 'We live in a world where deliberately dimming the fucking sun might be less risky than not doing it.'

The climate activist Greta Thunberg has compared the scale of humanity's task now, as it begins to confront, mitigate and redress the effects of two centuries' unregulated carbon dumping into the atmosphere, to the challenge that faced medieval stonemasons approaching the building of a cathedral. Those masons knew they'd never see the benefit of such cathedrals themselves, but worked with patience, reverence and ingenuity to build something they believed would benefit future generations. The medieval architects of cathedrals were often also the designers of bridges, and to me, those bridges might be a better metaphor for what we face.

As a child crossing the Tweed for a caravan holiday in England, my carbon footprint was negligible. At 18, I left the UK for the first time, on a bus to Prague. At 19 I went on an aeroplane for the first time, to Dublin – a flight so short that the pilot didn't reach cruising altitude before descending again to the runway. It was 1994, the Rio climate summit was two years gone, and the CO_2 in the atmosphere was just 359 ppm. It's arguable that the power of individual action is limited, but since 2017 I've been offsetting the carbon produced in the journeys I've taken, and pay a company to draw CO_2 out of the atmosphere on my behalf.* I have finally switched my home plumbing system to a renewables heat pump, and use an electric car. But I've a long way to go just to recapture the CO_2 released by the journeys described in this book.

The risks and costs of the engineering needed to undo the damage of centuries of atmospheric poisoning by human activities are high, but then perhaps the costs of Xerxes' pontoon bridge, or Li Chun's bridge over the Xiaohe, or Prague's Charles Bridge met with similar disbelief among the people of their age. Bridges require time, effort, money, energy, but afterwards we take them for granted. They are worth their costs in proportion to the extent to which they ease the lives of future generations. Life is a passage from one state to another, and we need to construct new ways of living that will grant safe passage to the future. *Bridges are good to think with.* Given the degree of warming already locked into the world's climate, we had better start building.

---

* There's a company based in Iceland that uses geothermal energy to turn atmospheric CO_2 into stone.

Chapter Seventeen

Bridge of Force

Murray, Australia
Hume Dam: concrete (1936), 300m
Borders: New South Wales–Victoria; Yorta Yorta–Jaitmatang

Mr Speaker, I move that today we honour the Indigenous peoples of this land, the oldest continuing cultures in human history . . . The time has now come for the nation to turn a new page in Australia's history by righting the wrongs of the past and so moving forward with confidence to the future.

KEVIN RUDD, AUSTRALIAN PRIME MINISTER

The bridge at Sydney Harbour is so charismatic and impressive that it has come to replace even the opera house of that distinguished skyline as an emblem of the city; a single-span arch of steel more than a kilometre in length. Sydney is a metropolis of five million people, and the monument at its gateway is as striking as any Statue of Liberty or Colossus of Rhodes. The decks of the bridge have a double use: they allow the passage of traffic as well as holding the arch in tension. There is no need in a bridge like Sydney Harbour's for towers at each end – the arch is self-sustaining. But we are accustomed to towers, and so all over the world tied-arch bridges have added unnecessary decoration.

Before Europeans arrived in Australia, Sydney Harbour was already a frontier of a kind. Though Aboriginal notions of territory and land ownership are different from European notions, the Australian Institute of Aboriginal and Torres Strait Islander Studies describes Sydney Harbour as the traditional frontier between the Eora people (who extended as far south as Botany Bay) and the Kuring-gai to the north. I paid a dutiful visit to Sydney's Harbour Bridge – to marvel at the scale of it, and climb to the crest of it – but I really wanted to see the bridges further west, along the Murray River, where New South Wales meets Victoria. An immense concrete dam there has forced the ecological transformation of an area as large as Ethiopia – the Murray Basin.

Many styles of bridge are designed to obstruct the flow of water. One of the most beautiful bridges in the world, the seventeenth-century Khaju Bridge in Isfahan, also functions as a dam, with sluices between its piers capable of creating a reservoir of water for irrigation and for cooling the city. The twentieth century has seen the proliferation of the most extreme examples of this kind of bridge: the concrete dam. The Hume Dam over the Murray River, 500 kilometres southwest of Sydney, took 17 years to build, starting at the close of the First World War and completed in 1936. It is named for a British traveller who began his journeys into Australia's vast interior with Aboriginal guides who subsequently abandoned him, and though it's just 300 metres or so across, that narrow pinch of cement has flooded 6,000 square miles of territory. With those few yards of a dam, the power, force and possibility of reinforced concrete is made manifest.

E. and I were travelling from the Hume Dam in the east to Echuca in the west, following the flow of the Murray River. Our journey was made with a group from Melbourne's Friends of the Earth, who were taking a visitor from Papua New Guinea on a tour of Aboriginal communities. Mel had a black felt of a beard, eyes like chips of ore and a slow, musical accent in English. He told me that land in his own country was being bought up by Chinese mining companies who sucked out minerals and money, under licence from PNG's government, but left only slag heaps and poisoned rivers. He was keen to learn ways of resistance to the exploitation of his native land from the Aboriginal communities along the path of our journey.

Rainfall is so precious in Australia that the Hume Dam provides water for towns and cities as far away as Adelaide, almost a thousand kilometres away, making life possible for two million people or more. But it has reversed the water cycle of Australia's largest river, which naturally runs high in winter and spring but now runs highest in summer, when the sluice gates are opened to better irrigate the fields. More than 50 per cent of Australia's agricultural production is reliant on the dam. The water exiting from its underside is several degrees colder than it would naturally be, with devastating effects on the kinds of life it can support. NASA has calculated that the volume of water that human beings have locked up behind dams is now so immense that it has slowed the rotation of the earth.

In 1770 Captain James Cook claimed the east of Australia for the British crown. Eighteen years later it became a penal colony, and the expropriation of land began. Many Aboriginal people who resisted were killed – few of those massacres are recorded. In 1837 a London *Report of the Parliamentary Select Committee on Aboriginal Tribes* took a dim view of the killing by British colonials of indigenous peoples – whether they be

in Newfoundland or South Africa or Australia. 'Their claims, whether as sovereigns or proprietors of the soil, have been utterly disregarded,' it pointed out. 'The land has been taken from them without the assertion of any other title than that of superior force.'

The authors of the report deplored the violence, but not the colonialism: 'In the recollection of many living men every part of this territory was the undisputed property of the Aborigines, it is demanding little indeed on their behalf to require that no expenditure should be withheld which can be incurred judiciously for the maintenance of missionaries, who should be employed to instruct the tribes, and of protectors, whose duty it should be to defend them.' They concluded that in compensation for having their land stolen, the indigenous peoples throughout the domains of the expanding British Empire should be compensated not with restitution of their lands, but with a reliable supply of missionaries and soldiers.

Following the death of Queen Victoria in 1901, Australia shifted from having the status of a colony to being a federal state under the new king. The previous year Henry C. Morris had written in his *History of Colonization*: 'If the workforce of a colony cannot be disciplined into producing the profits rightly expected by the mother country the natives must then be exterminated or reduced to such numbers as to be readily controlled.' His near-contemporary, George Pitt-Rivers, wrote in 1927:

> The survival of the natives will only cause trouble . . . There is no native problem in Tasmania, and for the European population in Australia, the problem is negligible, for the very good reason that the Tasmanians are no longer alive to create a problem, while the aboriginals of Australia are rapidly following them along the road to extinction.

In 1936, the Native Administration Act of Australia proposed a legal route by which Aboriginal people were to be 'bred out'; this was chillingly described as a 'final solution' to the problem of race in Western Australia. The following year a day of mourning was called by Aboriginal peoples to mark 150 years since the establishment of Australia as a penal colony under the British crown.

A referendum in 1967 returned the verdict that 90 per cent of registered voters agreed that Aboriginal Australians should be included in the national census – before then, indigenous peoples were considered less than human. The right to vote in national elections didn't follow until 1976. In 1991 the prime minister of Australia, Paul Keating, conceded that 'it was we who did the dispossessing. We committed those murders.'

In February 2008 I was in the Western Australian town of Norseman when the then prime minister, Kevin Rudd, issued a further acknowledgement and apology, later called 'Sorry Day', focused in particular on the forced abduction of children from Aboriginal communities. I watched his speech on an open-air campsite television, surrounded by white Australians on holiday: 'We apologise for the laws and policies of successive Parliaments and governments that have inflicted profound grief, suffering and loss on these our fellow Australians. We apologise especially for the removal of Aboriginal and Torres Strait Islander children from their families, their communities and their country.'

It was the beginning of some kind of attempt at restitution, though its foundations are still fragile. The Hume Dam took seventeen years to build; Sydney Harbour bridge took nine – by contrast it will take many decades to build bridges of reconciliation between settler and indigenous communities in Australia.

Back at the Hume Dam I watched trucks thundering past and water roiling up from its foot, and for a moment the force of human ingenuity seemed limitless. The sun was hammering down its gavel of light and heat,

a sun that's feeling hotter every year. The water was cool, it was life, and the holding back of the river had brought more life into a vast area of southeast Australia. But was it the right kind of life?

The climate crisis is playing out with particular ferocity in Australia. The damage wrought by the dam showed that human ingenuity *does* have limits – we cause harm with every intervention, and our attempts at fixing our mistakes often make things worse.

The line of the Murray River marks a boundary of the Jaitmatang nation; on the banks of the river Mel and I were introduced to a woman of that heritage, who told us of the ways her ancestors had traditionally crossed the water. They'd find a trunk of eucalyptus of about the right girth for a canoe, and carefully strip its bark in one long sheet, curved at bow and stern. The bark was then shaped into a boat. A nearby tree she indicated had a long ellipse of a scar; in the 150 years since it was first cut the bark had rebounded, like an eyelid closing slowly over an eye.

Further downstream, the Murray slides between lands once dominated by nations known as Yorta Yorta and Baraba Baraba. We were introduced to an elder of the former, who began an account of the ways land had been swindled from his ancestors. The land we stood on had never been ceded, he said, but was still illegally occupied. It had been taken by immigrants from Europe, who had snatched it by force of arms, using laws that had no jurisdiction in the country they occupied.

The Yorta Yorta had taken their claims to an Australian court, but been told that their claim had lapsed, not through some flaw in their argument, but because of the 'tide of history' – history being written by the victors.

Two years after I left Australia, in 2010, the National Congress of Australia's First Peoples was established, and in October 2023, after ten years of campaigning, a referendum was held on changing Australia's constitution to give Aboriginal Australians a greater voice in parliament. Sixty per cent of the population voted 'no'.

Just as Mel had come from Papua New Guinea to learn from Aboriginal nations, the Yorta Yorta had learned from Canadian First Nations – they were creating maps of stories and memories of their people, before such memories were forever washed away. The trauma of colonialism is forging unexpected alliances, and it's not yet clear what kind of new bridges between peoples will emerge.

2010s

Chapter Eighteen

Bridge of Home

Forth, Scotland
Queensferry Crossing: cable-stay (2017), 2,700m
Border: Lothian–Fife

She erected dwellings on either shore of the sea which divides Lothian from Scotland, so that the poor people and the pilgrims might shelter there and rest themselves after the fatigues of their journey . . . Moreover, she provided ships for the transport of these pilgrims both coming and going, nor was it lawful to demand any fee for the passage from those who were crossing.

BISHOP TURGOT OF ST ANDREWS,
LIFE OF ST MARGARET

The view west from the Forth Road Bridge is no less remarkable than the view towards the rail bridge in the east: the estuary narrows as it reaches towards Kincardine and Bannockburn at Scotland's waist, and at sunset the north-western horizon is a sawtooth silhouette of hills. In 2010 I moved to Queensferry from Orkney, and a year later work began on a new bridge that would interrupt that western skyline of the Forth – the Queensferry Crossing. In the three years since travelling in Australia I had become a father of three. It was my turn, now, to endlessly read the Billy Goats Gruff with a child on each knee; my turn to trace my finger slowly over the Ladybird book of bridges, showing how shapes of ink on the page are bridges to sounds, and those sounds are intimately and transformatively connected to ideas.

In its name, the new bridge recalls those medieval boats that put out at Queen Margaret's command in the eleventh century. It was named by public vote from a shortlist of five – alternatives included 'the Caledonia Crossing' (recalling that Roman frontier) and 'Queen Margaret's Crossing' (to commemorate the founder of the free passage over the water). To the inhabitants of medieval Queensferry the decks of the immense bridge now standing over the water would have seemed miraculous – a passageway through the sky built by a confederation drawn from across the former Roman Empire and beyond.

One morning during its construction, before the north and central decks were joined, I took a tour of the Queensferry Crossing. I'd been asked to write something about it for BBC Radio, and it was a treat to have the chance of eliding that building of the imagination with something more tangible. I met my guide for the day at the crossing's temporary offices on the north shore where we suited up in high-vis jackets and hard hats, and read the day's weather forecast ('chance of high winds', it said, 'squally showers').

We passed workers from Scotland, England, Germany and Latin America as we took a rough path up the northern embankment towards a scaffolding staircase. The stair emerged onto the north deck where thousands of tonnes of concrete were being poured into a mesh of rebars – the rough steel rods that lend the structure immense strength. The casings of the cables were the most striking sight – they could be seen plunging from the towers overhead into sockets moulded through the decks, each fixing as broad and strong as an anchor chain for the *Titanic*. The casings protected hundreds of cables, and the individual cables were wax-coated then sealed in rubber to avoid the corrosion problems that have plagued the adjacent road bridge.

Standing on a place in space that had been mid-air just weeks before, I looked over towards the bridge I had run across for charity back in 1988. My guide pointed out the fittings where wind-shielding would be installed – the new crossing would remain open in almost all weathers. Cables would later be strung between them so that would-be suicides couldn't

clamber through. The towers themselves were hollow, and I climbed inside to get a feel for the strength of their walls, dense with almost as much steel as they are with concrete. They stand 210 metres tall, the highest bridge towers in the UK.* Inside was a trabecular network of stairwells and pulleys, like the lattice of marrow down the core of a bone – reminding me that bone is stronger than concrete, though lighter. Leaving the tower we teetered out towards the edge at the free end of the deck, where workers were preparing to lift a new segment from a barge on the estuary below. From a height of over two hundred feet the surface of the water looked flecked with tufts of cirrus.

The steel base-plates of each bridge section were cast in Shanghai then floated to the Forth on ships, before being lined up along the Fife shore to be cast in concrete. Once set they were floated out onto the estuary on

* Though diminutive in comparison with the Millau Viaduct in France, which reaches 345 metres.

barges before being hoisted into position – it was necessary to raise the deck segments two at a time, one to each side of each tower, to balance the forces pulling down on the towers.

Politicians left, right and centre, conscious of Scotland's engineering and industrial history, were desperate to source the bridge's steel closer to home, but only Shanghai had the capacity to deliver the quantity required on schedule – which they did in late 2013, a few months before Scotland held a referendum on independence.

As the country voted on whether to leave the UK, the Queensferry Crossing's towers had risen above the waterline and reached the height of the future road deck. They wouldn't reach their final height until a year later, September 2015. The outcome of that referendum was a vote of 55:45 to stick with the status quo and remain in the United Kingdom.

Though much of the day-to-day building work involved Morrison's, a Scottish company, its towers were planned by a German firm, its approaches built by a Spanish, its welding done by Croatians, while the cable-stay and deck assemblies have been overseen by Americans. I heard how, in the lead-up to the independence referendum, there was pressure on the bridge contractors to emphasise the Scottish element of their workforce. In the months prior to the EU referendum, they were asked to pivot, and emphasise instead those workers who had come from Germany, Croatia and Spain. When I first heard this story I groaned at the cynicism of electoral politics, but perhaps all great works are inseparable from the patronage that make them possible.

By the time construction work had finished the new bridge had incorporated 23,000 miles of cabling, 35,000 tonnes of steel, 150,000 tonnes of concrete. Its three columns reached 700 feet in height, each one radiating a spray of silvery-white cables, like the outstretched wings of three Angels of the North. Sinking the caissons of the south tower through 35 metres of seawater and sediments was an unprecedented task, and one of the most challenging of the project: the silts had to be scrubbed back to clean rock and 17,000 cubic metres of wet-setting concrete poured in – a new world record.

From the north windows of my house I had watched the weight-bearing towers rise, each flanked by one of scaffolding – they resembled the service structures of a Soyuz or a Saturn V. Once I watched an aircraft carrier edge under its decks. So immense was the ship, constructed upstream, that it had to pass at low tide, and even then cleared the bridge only by inches. The cables are static but lend the towers the illusion of movement, as if each one is straining upwards but held back, a Gulliver pinioned by the ropes of Lilliputians.

In 2016, as the bridge neared completion, the UK held its referendum on membership of the EU. The final result of that vote was 52:48 to leave (though in Scotland it went the other way, at 62:38 to stay). The vote took place a couple of months before the central tower and its decks became the largest free-standing balanced cantilever ever built.

At that moment, just before closure of the central and southern segments, a gap of a mere six metres separated the two decks – a length that could easily be leapt should a long-jumper feel brave enough to try. During every storm that passed, the hands of the wind tested the bridge for weaknesses, but none were found.

As the UK voted on separation from its neighbours across a land border in Ireland, and through the Channel Tunnel in France, I had the opportunity to sail back and forward under the decks of this new icon of connection – feeling like plankton sieved from the mouth of a whale. The scale of its construction was difficult to grasp from ground or sea – the closest I've come to appreciating the bridge in context is on flights to and from Edinburgh Airport, when on take-off or landing you see it from above as a silver thread stitching the coast of Lothian to Fife. When you pass the bridge end-on, and look along its length, those cables resolve into a tapered line as if you're staring along the edge of a honed blade. On windy days its towers seem like oars pulling the earth through the sea of the sky; every so often they tear open a cloud and spill great arches of rainbows.

I've watched often from the Queensferry shore: halogen lamps by night illuminate the cables into a static shower of shooting stars; the towers are lit sparklers. Brief flashes of traffic headlamps break through a line of horizontal deck lights, as if the bridge is sending messages in the dots and dashes of Morse code. Following warm days, haar eases in from the North Sea and swallows the decks, leaving the towers floating as if on beds of vapour. But the eeriest view is when clouds creep slowly down the towers, granting a vision of the unsupported decks hovering magically over the Firth of Forth.

The possibilities of engineering seem endless; the cost of the bridge's construction came in on budget at £1.4 billion, or about £280 for every citizen of Scotland. The success of the bridge, and the economic savings it allows, poses a perennial question. What other bridges should we now be building?

Chapter Nineteen

Bridge of Balance

Wang Chhu, Bhutan
Thimphu Old Wooden Bridge: timber cantilever (approx. 1650), 55m
Borders: China–Bhutan; Bhutan–India

In perfect balance, searching of the Norm,
Perfect in knowledge and good practices,
Perfect in concentration of your thoughts,
Ye will strike off this multitude of woes.

DHAMMAPADA, 10:144

At the Junction bookshop in Thimphu the manager was reading *The Age of Reason* by Jean-Paul Sartre. 'I've been trying to get hold of *Nausea* for months,' she told me, 'but the Indian distributors won't send it up.' On a stand in the centre of the shop there were glossy photo books: cute, scruffy waifs; austere Himalayan panoramas; a coffee-table celebration of carved wooden phalluses (the Bhutanese strain of Buddhism employs phallic symbolism with zeal). These were the books laid out for souvenir shoppers. On the shelves, there was a section dedicated to Ancient Greek drama, another to nineteenth-century Russian novelists (all in English translation). There was a volume of Elizabeth Bishop and several volumes of Freud. The bookseller had sold her last copy of David Foster Wallace's *Infinite Jest*, but still had a copy of *The Pale King*.

I took a copy of Barthes' *Mythologies* over to the counter. On the floor was a stray dog, one of the custard-coloured mongrels that roll in Thimphu's dirt by day and howl to one another at night. The manager stroked the dog's patchy fur. 'His name is Motay,' her companion told me. 'It means "the fat one". People here feed him because he barks only at the police.'

On the main square outside there were monks and nuns wearing burgundy robes; some had prayer wheels, others had cellphones. Most of the local men were wearing the *gho*, a robe with a knee-length skirt a little like a kilt, and the women the ankle-length *kira*. Bhutan has been adamant that its traditional dress be more than a gimmick for tourists: at many of the city's institutions there were signs insisting 'Formal Dress Only'.

I sat down with the Barthes and opened to 'The Lost Continent', an essay that scolds the West for stereotyping and exoticising the East. 'This same Orient which has today become the centre of the world,' Barthes wrote, 'we see . . . all flattened, made smooth and gaudily coloured like an old-fashioned postcard.'

Not long after returning home from Australia I'd published my first book, and this invitation to Bhutan had come because of my second, *Empire Antarctica* – an account of living in Antarctica as an expedition doctor, alongside a colony of 60,000 emperor penguins. Thanks to its publication I was beginning to learn something of a worldwide community of letters, and experience in a modest way the reach of the literary world. The Bhutanese literary festival Mountain Echoes was being advertised by sky-blue posters of the Buddha. Some had been fixed to the walls outside the shop, and above a banner of sponsors' logos, the Master of Dharma was quoted as saying: 'If You Want To Find Wisdom, Throw Away Your Books'.

Book-chucking seemed an inadvisable way to promote a literary festival (particularly on the walls of a bookshop) but perhaps the slogan was intended as a koan, the sound of one publicist's hand clapping. Bhutan seemed a place of paradox; it had freedom of speech, but cigarettes were illegal. Capitalism had been introduced by an absolute monarchy, which had given way to a constitutional democracy less than a decade before. In its library I saw thangkas and prayer scrolls, intricate as maps but more jewelled and precious, and with the dignity of a still-living faith. In the West, Bhutan is largely known for its much-vaunted neglect of GNP in favour of GNH (Gross National Happiness), as well as the forced exodus, between twenty and thirty years ago, of more than a hundred thousand Hindu migrant families to Nepal. At one of the discussion events, a festival director, Namita Gohkale, described Bhutan as 'the one country in the Himalaya that has the privilege of maintaining its culture and its traditions'.

The walls of Thimphu Valley are steep and smoothly contoured, like a half-open almanac. Along the seam of it runs the Wang Chhu which, like many rivers, changes name as it crosses an international border – in India it is known as the Raidāk. A tributary of the Brahmaputra, the Wang Chhu makes its way from the Great Himalaya at the Tibetan border down to the narrow Siliguri corridor – a strip of Indian territory that arches over Bangladesh from the Ganges plain towards its seven northeastern states. I'd long wanted to visit one of those states, Meghalaya, where bridges of living tree roots have for centuries been woven over steamy tropical tributaries of the Brahmaputra. Unlike the rope bridges of Peru those root

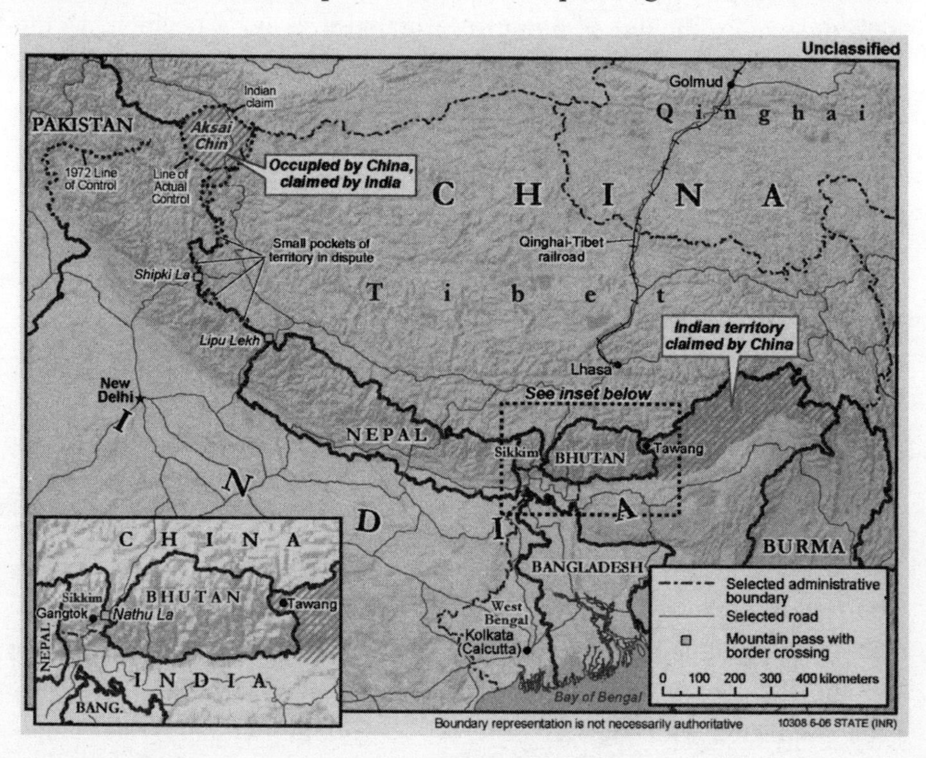

bridges are immune to rot, as long as they are well tended – as good a metaphor for the bridges between peoples as I'd likely encounter anywhere in the world.

Cantilever bridges of wood flourished in Asia while Europe was wedded to arched bridges in stone. In the seventeenth century, as Galileo's mathematics was turning engineering into a precision science, far away in Bhutan a balanced bridge was built over one of the tributaries of the Brahmaputra. In Thimphu I wanted to see the oldest bridge in town: cantilevered lengths of timber braced into a stone tower on each bank, upon which is balanced a trussed walkway. The whole bridge is roofed over to keep the rain off the vulnerable (and at those altitudes, precious) wood.

I approached from the town centre, on the west bank of the river. Squat towers like pillar boxes faced one another across the water. A boxy wooden cage rested on dark beams draped with prayer flags. The flags were so thick that the pedestrians crossing the bridge were almost obscured by them. I looked down on the gravel in the riverbed, glinting like coins in a wishing well. Up from the water came a sound, the voice of breaking stones, the rumble and creak of a riverbed scoured clean by glacial melt. An Italian bridge engineer, Enrico Tubaldi, once explained to me how the amplified rainfall occasioned by climate change is accelerating the wearing away or 'scour' of bridge foundations, all around the world.

An earlier iteration of the Thimphu bridge was over three centuries old when it was washed away by floods in 1958, and the expertise to rebuild a flood-proof version wasn't available locally. Its recent reconstruction was instead led by a team from the Swiss Technical University – the Swiss being perhaps the best-placed nation to advise on flood-proof mountain bridges over glacial meltwater, in a time of rapid global warming.

Post-colonial geography is often absurdly inconvenient, but Bhutan's position teetering over the Siliguri corridor is particularly so. Since the Dalai Lama fled to India in 1959 Bhutan has had no diplomatic relations with Beijing. The border to its north was closed then, and foreign relations have been conducted with the assistance of New Delhi ever since. The Indian military watches the Siliguri corridor with anxiety, referring to it as 'the chicken's throat' because of the ease with which it could be grasped (and wrung) by the Chinese army.

In its construction and materials, the bridge over the Wang Chhu was different from all the bridges I was accustomed to. It was the oldest I'd visited that had been built in timber using only traditional, pre-industrial methods. It seemed appropriate that it was in Bhutan, a place committed to the preservation of tradition, that I was to encounter it. The bridge was in harmony with the valley on either side of it, balanced on its cantilevers across the swift-flowing Himalayan meltwater. For centuries the kingdom of Bhutan has been a bridge between Tibet and India, but with the border to its north closed, it has been cut off from the free flow of cultures that have traditionally nourished it.

At the festival there were discussions, screenings and lectures on Himalayan culture and heritage from both Indian and Bhutanese perspectives, as well as debates about preserving Bhutan's traditions against its extraordinary pace of development – which is striking even by South and East Asian standards. In the late 1950s, when Barthes wrote *Mythologies*, there were still no roads in the country (outsiders arrived by mule, on foot or, if they were Indian diplomats, by helicopter). Television and internet were introduced a little over a decade ago, and the first freely elected prime minister took office in 2008, making Bhutan the world's youngest democracy.

To assess the properties of a material, chemists study the way it interacts with others. In Bhutan, along the high Himalayan frontier between Asia's two superpowers, potent reagents are being thrown into the mix: Buddhist theocracy is fizzing with free-market economics; Chinese inducements are starting to abrade India's privilege; donkey tracks are being poured with tarmac; the world's happiest people are learning how to want.

But if the bookstore was anything to go by, what some of them seemed to want was Freud, Barthes and shelf-loads of Tolstoy. Books are the bridges that promote understanding between cultures, and a way into an appreciation of other lives and other minds. Poised on the stone wall of the Himalayas, the Bhutanese wanted information that might help them tiptoe along a balance beam between Chinese and Indian ambitions, and emerge with their pride and culture intact.

Chapter Twenty

Bridge of Immensity

Yenisei, Russia
Krasnoyarsk Trans-Siberian Rail Bridge: steel truss (1899), 907m
Border: eastern–western Siberia

Bare banks, canals without quays, bridges everywhere testified to the recent triumph of the human will over the hostile elements.

ALEXANDER PUSHKIN, *PROSE STORIES*

Almost thirty years ago, as a medical student, I saw a then-novel type of IVF performed. Under magnification a tiny hollow needle was advanced on a human egg. The needle was invisibly fine to the naked eye, but under the microscope seemed more like a cudgel. It advanced slowly, indenting the cellular wall of the egg, building in pressure, until suddenly – silently – it entered the infinitesimal. The DNA of a sperm was injected directly into an ovum.

Advancing into a new culture, language and landscape feels a little like that moment. You arrive in a new place, your home atmosphere carried with you as close to you as your own cytoplasm, and at first you remain intact even as the new world presses in on you. It pushes in on your skin, into your mind, your thoughts, your dreams. And then suddenly – a rupture, and a new fertilisation begins.

An invitation came through for me in 2017 to attend a Siberian book festival and speak with the medical students of a Siberian university. Both venues were in Krasnoyarsk, on the banks of the River Yenisei, or Енисей, the central of three immense rivers that traverse Siberia, and one of my favourites to trace in my many hours of atlas-wandering. My invitation was from the British Council of Moscow, and my visit was to make a bridge between cultures – literary and medical, British and Russian. After calculating the cost of offsetting the carbon, I signed up. The Yenisei! It wasn't possible for me to refuse.

Krasnoyarsk Oblast: five hours of flight over conifer and snow, tracks cut through the forest leading to illuminated petrochemical installations. The city I was aiming for lies at 56° north, the same latitude as Edinburgh, but on a longitude of 90° – quarter the span of the globe away from Scotland, and as far east as Bangladesh. Having flown that great distance, we were still only halfway across Siberia.

Arrival was in darkness, into wind-blown hail. Broad streets without pavements; through one window, a small hairdresser's shop, I saw a wall papered with a tropical jungle scene, so lush and inviting in comparison with the scene outside – of pinched faces beneath fur hats, and hurried poverty. There were Krispy Kreme shops and Costa Coffee, and my taxi driver told me with pride that there would soon be a McDonald's. Car showrooms, vehicle graveyards, mud-caked Soviet-era trucks hauling iron-sealed skips. Beyond the ramparts of the valley horizon I knew the road continued on through a forest vastness to reach Irkutsk and Lake Baikal – its waters a tributary of the Yenisei and the oldest, deepest, largest by volume and clearest lake in all the world. My appreciation of Cyrillic extended to the road signs, but little else: 'Novosibirsk 810km', 'Irkutsk 1,101km'.

From my hotel window I watched traffic passing over the Yenisei River from western Siberia to east. It is the fifth-longest river in the world, arising in the hills of Mongolia and flowing north like a compass needle all the way to the Arctic. When the Japanese and the Germans decided to divide Asia between them they chose the Yenisei as the border, their frontier and connection between the master races.

The wind on the bridge carried particles of frost that rasped my skin like sandpaper – I had to pull up my collar and walk sideways onto it as it came shrieking off the river. Ice began to form in my beard as I crossed, and tugged at the skin of my face as I grinned.

At the hilltop sanctuary of St Paraskeva I lit a candle as a mark of respect, perhaps for the sufferings of the people of Siberia. When the writer Colin Thubron passed through Krasnoyarsk in the late 1990s he wrote of how the western suburbs of the city were advancing into the rolling forest, the *taiga*, and beneath him the city was a pool of slums, smokestacks, cloaked in the pall of dirty fires. Communism has been easily as rapacious as capitalism in terms of environmental destruction. To the southeast Thubron could almost make out the blur of factories and military installations that had made this a closed city in the years of the Cold War. Hunched hills in the distance were like the khaki shoulders of a marching army.

I arrived almost two decades later, and there was less pall. There were, though, the same timber houses tottered with tilted chimneys, their roofing felt ripped, their beams sagging, their windows broken; I realised with a shock that people were still living behind their beautifully carved if rotting facades. The city itself seemed in mourning, a straggling survivor of a war. A murder of crows surfed on the wind, a babushka sold icons in the chapel stall, a solitary dog-walker veered across the square to avoid me.

Krasnoyarsk was founded in 1628 by Cossacks who'd been charged by the Romanov rulers of Russia with extending their rule from the Urals to the Pacific. Its name means 'Red Steep River Bank', a direct Russian translation of the indigenous name for the place, Kyzyl Char, descriptive of the mud walls that rein in the river. Against the levee of that river was moored a steamship with bunk rooms that had been occupied by both Tsar Nicholas and Lenin. The tsar had been on a tour of his vast domains; Lenin was being exiled upstream to Shushenskoye – a village closer to Shanghai than to Moscow – where he would read and reread the books that would inspire and guide his revolution.

The great rail bridge of the Yenisei was designed by Lavr Proskuryakov, a man who has been described as a Pushkin of engineering, a poet in steel. His bridge connected the eastern and western stretches of the Trans-Siberian line, and was a winner at the Paris Exposition of 1900.

The prize was awarded by Gustave Eiffel, who, fifteen years earlier, had shown with his Garabit viaduct just what trussed spans of steel could do. The Yenisei railway bridge opened in 1899, six half-moon spans of trussed arches, in total over 3,000 feet long, among the longest bridges then in Asia. Its spans of steel lasted a century: through the 1990s they were progressively replaced with new trussed spans laid on the original piers.

In Krasnoyarsk's railway station I had a glimpse of the immense reach of this country, from the Gulf of Finland to the Strait of Tartary, and dreamed of hopping on board a train bound for the east. The train from Moscow was, I noticed, running 42 minutes late – not a bad slippage in the timetable considering it had left the capital two days before. It would be another couple of days before it reached the Pacific.

The bridge over the Yenisei I walked most often was just downstream of Proskuryakov's Trans-Siberian railway bridge. Its shallow concrete arches were raised high enough to survive floods, built a couple of years before the Cuban Missile Crisis. Flanking the river was a giant's staircase of levees, the broad concrete steps decorated with statues of water nymphs and Russian poets. The poets were better clad than the nymphs. I had with me some printouts of old postcards of the original Trans-Siberian bridge, and tried to find the best angle to reproduce the framing myself, but the Cold War development of the city had subsumed the river's banks and rendered them unrecognisable.

A handful of proud parents stood on the quayside watching their children give shrill, amplified speeches about the centenary of the Russian Revolution. Their numbers were not many. My guide around the city thought it disappointing that the centenary was not more widely commemorated. She told me that people were uneasy about whether to celebrate it or not. Was Putin for the revolution or against it? Was it something to forget or to applaud? Putin had recently unveiled a gallery of busts of leaders of the Russian people in which he played a starring role, the culmination of a line that ran uninterrupted from the tsars through Lenin, Stalin, Khrushchev, Brezhnev, Andropov, Chernenko, Gorbachev, Yeltsin, before arriving at himself. Seventy years of communism dissolved in a narrative of the glorious progress of Russian leadership.

The cultural centre was draped in billowing netting, as if in preparation for a demolition that had at the last minute been cancelled. There were installations on the wall – a man in a wheelchair launching a plane; a Necker cube dangling over a plaza, where free-standing door frames created their own sense of borders in space.

At the State Medical University I gave the talk I had travelled so far to deliver, about the universal utility of medicine – a profession needed everywhere from Africa to Antarctica, India to Indiana. Recurrent among the questions afterwards were the avenues for Russians to seek work abroad, given the visa restrictions they faced and the low pay and status of doctors in Russia. A pair of exchange students from Hokkaido had travelled half the breadth of Asia to build connections with their Russian neighbours; one of them told me that she hadn't realised it was possible, as a physician, to be more interested in people than in diseases, and that the practice of medicine might even benefit from greater engagement with art or with literature.

I tried to explain my own view: that only part of medical practice was its science, while another part, no less essential, was concerned with our common humanity, which extends across all borders. That medicine in its best manifestations made connections between ideas, helping patients to escape silos of suffering; that the arts could be an enlightening element of a doctor's training. Then we were ushered out of the marble corridors by a burly but avuncular security guard, and it was time to wrap my scarf and jacket tightly around myself, back through the storms of hail to the hotel.

One of the most impressive concrete dams in the world lies a few miles upstream of Krasnoyarsk, and I wanted to take advantage of my time in the city to see it. My guide at the university arranged a taxi, and the following morning we set off. The Soviet authorities of the 1950s intended to build the dam nearer the city, but the soil was too soft to take the immense abutments such a quantity of water would need for its support. It took sixteen years to build, 1956–72. Nearly all the energy it generates goes to supply an aluminium plant outside the city, fed by bauxite mined two hundred kilometres away. The dam's monstrous turbines were forged in Leningrad and shipped out of the Baltic, up and around the north cape of Europe, along the fringes of the Arctic Ocean, then brought up the Yenisei into this heart of Asia.

I stood at the base of the dam, thinking of little dams that as a child I'd built of turf; how the crest of every dam makes a bridge; how every dam represents the drowning of a river. On the downstream face was a boat lift, a giant bath on rails, to hoist river traffic up that wall of cement. Fast-flowing water emerged from its foot black as bitumen, smooth and uniform, relentless as a conveyor belt.

Just south of the dam I stopped at a village built for the dam workers, and where even today they are allocated housing. Rows of the ubiquitous five-storey apartment blocks lay along the riverbank – named Khrushchev houses, because it was under his rule that they were constructed the length and breadth of the Soviet empire. A monument to the town was an upturned V; I heard it called a 'monument for the manipulation of the people'. The point of the gulags was to make people work, it was said, telling them they *had* to work because they were bad. But the point of settlements like Divnogorsk was to make people work because they were good, they were heroes, they were the strength and the muscle of the Soviet Union, and should be proud even if that meant they had to survive the Siberian winter in tents. Behind the upturned 'V' was a tank – 'Always the army, in this country' – and beyond that, the Yenisei flowing seamlessly in its glacial trench.

'Look, there is no industry!' my guide said. 'It is so clean! They step out of their houses and they are in the forest, finding mushrooms!'

In those forests west of Krasnoyarsk, walking for hours in woodland so rich in silence it was as if all the trees were holding their breath, tiny tame nuthatches came to my hand – accustomed to taking seeds from the day trippers. The trees were grey and winter-thin, and made the distant hills look as if they were draped in gauze. Large warning signs informed me to guard myself against the depredations of bears.

Moscow from the air seemed a great spider's web of roads, each filament traced by the light of sodium lamps into a pattern of irregular polygonal circles and radial spokes, of mean Dostoevskyan streets and palatial Tolstoyan boulevards. As the plane banked over the city I followed the pattern of diminishing circles to its heart, and realised I was looking down on the Kremlin. Cocooned there at the centre of the web, I knew, was the waxy body of Lenin.

The November snow didn't seem to settle much on the birches in Moscow; their boles were pruned, their white bark raw in places, the branches snapped like bones. Roads were built wide enough for tanks to roll six abreast. As I passed through the city its blocks of apartments seemed interminable, punctuated only by diminutive onion-dome churches and occasional statues of heroes. It was dawn: amber furnace light glowed on the tree trunks that shimmered in the cold. I didn't expect so many hills in Moscow, so many river tributaries or so many bridges. I saw posters advertising theatrical productions of *Anna Karenina* and *The Cherry Orchard*, and concerts of Tchaikovsky and Mussorgsky. At one junction, from my low-slung seat in the back of a taxi, I looked out at hip height to see a gang of little boys, eight or nine years old, sauntering along the pavement smoking cigars.

The British Council had asked if on my one-day stopover I'd give three lectures. About books at the Gorsky Institute of World Literature, about medicine at the Timiryazev State Biology Museum and about approaches to the medical humanities at the Nekrasov Library. Each talk was intended to build connections between Britain and Russia.

The Gorky Institute is housed in a nineteenth-century mansion, built after the razing of the city by Napoleon's disordered troops. Bare birches grew in the garden. My host, Elena, met me with great generosity, warmth and smiles, on stairs flanked by columns that seemed designed more for the grand entrance of debutantes than for welcoming scruffy Scotsmen. She told me that she had studied Wordsworth at the University of South Carolina. 'This was Pushkin's daughter's house,' she added.

'I have never got very far with Pushkin,' I told her.

'Don't bother to try,' she replied. 'His poetry is untranslatable.'

The homes of elite Russians such as the Pushkins were collectivised after the revolution, and many were given over to academic institutions. But a counter-revolution was under way and, indicating the peeling paint of the ceiling and the blooms of mould on the walls, Elena told me that there was a rumour that the old academic institutions were to be re-appropriated by the state and handed over to banks, corporations and oligarchs. No one else had the funds to maintain them.

The literature and translation students listened politely as I spoke about my books, and asked me about life, death and regrets. One young woman wanted to know about the cemetery that every doctor carries in the mind, where they go to remember the many patients they didn't help. One young man, whose father was a neurosurgeon, told me that witnessing the appalling burdens of the man's vocation first-hand had convinced him to follow instead a career teaching literature. Another student, a scholar of Kenneth Grahame, told me how much he hated Moscow and wished to return to the simple village life so glorified by Tolstoy. Tolstoy was too moralistic and too in love with humanity, another said; Dostoevsky was too harsh and unforgiving said another.

At the biological museum, the questions from the audience were all on Brexit, and life in Western Europe. Why, they wanted to know, did Europe have a migrant crisis? There was no such crisis in Russia, because the police were very strong. Putin was perhaps too interested in Moscow at the expense of the regions of Russia, but without him the regions might break off from the state. A table of Johnnie Walker was laid out, and discussions of migration, politics and medicine carried on lubricated by the drinking of Scotch.

It was late afternoon when, heavy with whisky and a lunch of omul – fish from Lake Baikal – I stumbled towards the Nekrasov Library. Questions there were about the value of literature to medicine, and the value of medicine to literature. I was asked to recommend Scottish poets who might match Nekrasov in their fluency. Whether trepanning your own skull might be good for your health, and whether robots might soon render the job of both writers and doctors obsolete. 'I don't think so,' I replied. 'I can't imagine medicine *or* literature practised without a human connection.'

In the departure lounge waiting for my flight home the following day I met Natasha, who was glad of the chance to practise her English. She explained to me that people in Russia were aware of the #MeToo campaign against sexual assault then dominant in the Western media, but didn't much like it. 'Some of the women only do it because they want others to think how attractive they are,' she said.

'Isn't that too harsh?' I said. 'Surely it is better if such behaviour is exposed?'

'Every woman has a story like this,' she said, 'but to shout out about it,' she frowned, 'it doesn't help.'

She asked me about Scotland, and told me she'd read in Russian newspapers that some people in Scotland were seeking independence from the UK – was this true? She also asked about Brexit and the UK's separation from the EU. I explained as best I could the positions that others had put forward, the competing tensions. 'Brexit will be a good thing for UK,' she said with finality, 'in about fifty years' time.'

'That's longer than I'm likely to live,' I replied. 'Politicians who risked nothing themselves, who have the freedom to live anywhere in Europe, voted to strip me of one of my most valuable liberties.'

Natasha waved her hand, dismissing my complaint; for her, diplomacy and cooperation had their limits. 'Tough luck for you,' she said. She was all for bridges between nations, as long as those bridges could be protected, even militarised, and were able to be blocked off at short notice.

Russia had annexed Crimea three years earlier, and our discussion of Scotland, UK and EU politics moved on to the subject of Ukraine. Her manner changed abruptly. 'I have family in Ukraine,' she said, 'and it's sad, but now we don't speak at all. Their nationalism has torn our family apart.' She told me that the word *u-kraina* means 'at the edge'. 'It is the edge of Russia, *it is Russian*,' she insisted, beating her leg with her fist.

In the years since that conversation – and especially since Russia and Ukraine have been at war – I've thought often of our conversation, and the unyielding way that Natasha beat her leg to dismiss Ukrainians' claims to independence. She then waved her hand as if to indicate the city, and beyond it the steppe, the forests of Russia, its mighty rivers, the trains trundling back and forward between Vladivostock and St Petersburg. 'The Ukrainians, they just have to realise they are part of *this*, they are part of *us*.'

Chapter Twenty-One

Bridge of Occupation

Jordan, Palestine
Allenby Bridge: steel truss (1918), 90m
King Hussein Bridge: concrete box girder (2001), 90m
Border: Jordan–occupied Palestinian territories

Then the Lord said to him, 'This is the land I promised on oath to Abraham, Isaac and Jacob when I said, "I will give it to your descendants." I have let you see it with your eyes, but you will not cross over into it.'

DEUTERONOMY 34

In a Starbucks in Amman I met Suad, a Palestinian woman whose family, she said, was from Ramallah, though she was born in the state of Jordan. She had very pale skin, an East Coast American accent, and earrings of silver and aquamarine that trembled as she spoke. In Boston she'd graduated with a master's degree in human rights law, and she now worked at the UN's Amman office interviewing incoming Iraqi refugees. Every day brought new salvos of gruelling, upsetting stories; her eyes filled with tears at their retelling. Many Christians, she said, had been raped by soldiers; she'd interviewed scores who'd spent time in the prisons of Abu Ghraib. Her job was to process refugees destined for the US, Australia, Sweden, the UK. She told me with weariness that on her desk were the details of seven thousand people she had to process by next month, for onward transit to America.

Some Palestinians in Jordan speak with gratitude about their position in the country, but not many; others told me the king was an enemy, that Queen Rania was a traitor for marrying him, that Jordanian Palestinians had become domesticated, that the 1960s generation of Palestinians were mad for still believing in Arab nationalism, that Jordan as a country was in denial about how much it owed to the wealth and industry of Palestinians, that their nation was itself a pathetic buffer state, its ludicrously angular borders drawn up by British and French diplomats on the back of an envelope.

Palestinians who fled east across the River Jordan in 1948 and again in 1967 make up more than 60 per cent of the population of Jordan. Many of those who fled used the Allenby Bridge – named for the British general whose forces built it during the First World War as part of their campaign to defeat Ottoman Turkey. Earlier in that campaign they had conquered Jerusalem. 'The importance of Jerusalem lay in its strategical position,' wrote Allenby later. 'There was no religious impulse in this campaign.' In taking the city he was at pains to avoid using imagery drawn from the medieval Crusades.

When the Romans occupied the Jordan Valley they too built bridges; they had a wooden crossing constructed at the same site of the later Allenby Bridge. The Ottomans had one here too, which they blew up in 1917 to slow the British advance. General Allenby simply had a new version made, its structure of interlocked triangles of steel, its base of wooden timbers.

The simple truss bridge built by Allenby's men was damaged by an earthquake in 1927, by Jewish fighters in 1946, and again in the Six-Day War of 1967. In the 1990s the Palestinian Authority asked for the Japanese government's help in replacing it. The result was the King Hussein Bridge, a short, sturdy concrete crossing of box-girder type, modest and replaceable, anonymous above a river that carries such a freight of stories, faith and myth.

In his book *I Saw Ramallah*, the Palestinian writer Mourid Barghouti wrote of the original Allenby crossing that it was more border than bridge:

How was this piece of dark wood able to distance a whole nation from its dreams: to prevent entire generations from taking their coffee in homes that were theirs? . . . I do not thank you, you short, unimportant bridge.

The day after meeting Suad, E. and I stood on the crest of the hills west of Amman at Madaba – Mount Nebo – where, according to the story recounted in the King James Bible, about three and a half thousand years ago Moses received the vision that gives his heirs such certainty of ownership over historic Palestine:

> Then Moses went up from the plains of Moab to Mount Nebo, to the top of Pisgah, which is across from Jericho. And the Lord showed him all the land of Gilead as far as Dan, all Naphtali and the land of Ephraim and Manasseh, all the land of Judah as far as the Western Sea, the South, and the plain of the Valley of Jericho, the city of palm trees, as far as Zoar.

There are some who insist this passage gives Israel rights over not just the occupied West Bank, but much of the land east of the Jordan. As soon as all this territory was promised to Moses, however, he was told by his God that he would never be permitted to enter it, but would die there east of the River Jordan: 'So Moses the servant of the Lord died there in the land of Moab.'

The Jordan plain stretching beneath us towards Jericho was opaque, beige in the dusty heat. A Byzantine mosaic in the floor of the old Christian church on the hilltop showed a River Jordan full of life, its banks lush and green – it was by tradition one of the four rivers of Eden. The mosaic even showed a ford used to cross between the east and west banks of the Jordan, the forerunner of Allenby's bridge. There was no border shown on the mosaic of the Jordan crossing; the land east and west of the river was indivisible, though the Old Testament stories often frame it as a border, or at least a military frontier:

> And the Gileadites took the fords of Jordan before the Ephraimites: and it was so, that when those Ephraimites which were escaped said, Let me go over; that the men of Gilead said unto him, Art thou an Ephraimite? If he said, Nay; then said they unto him, Say now Shibboleth: and he said Sibboleth: for he could not frame to pronounce it right. Then they took him, and slew him at the fords of Jordan: and there fell at that time of the Ephraimites forty and two thousand.

In 2022 there were reports that Ukrainian forces were demanding the same kind of life-or-death linguistic test of their prisoners: that each pronounce the word *palianytsia* (a type of bread). The way the word is said framed as a reliable indicator of Russian or Ukrainian upbringing.

The stretch of the River Jordan where Jesus was baptised has been hesitantly identified by archaeologists by tracing the remains of a Byzantine church. I walked towards it under fierce sunlight through a sparse grove of tamarisk trees, the heat like a ramshorn blast. Goliath was said to have been killed nearby; the Qur'an describes the Hebrew army resting under these tamarisk trees on their way to battle the Philistines.

The waters of the Jordan were sadly diminished, no more than a runnel. They trickled south as if having to win every yard against unspeakable odds. The river was not much bigger than the streams of my childhood games – and very far from the powerful cerulean flow shown on the Madaba mosaic. Its waters are siphoned off now for agriculture, the rains of the region lessened by climate change, vapour boiled off by the heat, so that, as often as not, its dribble no longer reaches the Dead Sea.

On the opposite, occupied bank was a rival Israeli baptism site, with platforms descending to the muddy western bank of the river. Soldiers faced one another across a few yards of stream, Jordanian and Israeli, both looking on in bemusement as tourists, mostly Russian, stepped down into the water to be washed clean of their sins. An Orthodox priest in robes of burgundy with golden tassels swung his censer, and cajoled me to join in with a Russian hymn.

Philistines / Palestinians – the names are echoes of one another. Through the haze of heat I could make out the foothills beneath the city of Jerusalem, perched on its scarp to the west of the border. I stood on the approach road for King Hussein's Bridge, but didn't go closer. It was heavily guarded, and because of the ongoing occupation it was impossible for me to cross: a stamp in my passport would deny me entry into many of the Muslim countries further east on my journey, because they reject Israel's illegal occupation of the Jordan's west bank.

Mourid Barghouti's memoir is remarkable for the humanity with which he describes the Israeli soldier who blocks his passage over the bridge – reflecting on whether the soldier serves Israel willingly or reluctantly, whether he was born in Israel or is a recent arrival from Brooklyn, whether his parents are Holocaust survivors, whether he kills for duty or pleasure. 'The Jordanians call it the King Hussein Bridge,' Barghouti wrote. 'The Palestinian Authority calls it al-Karama Crossing. The common people and the bus and taxi drivers call it the Allenby Bridge. My mother, and before her my grandmother and my father and my uncle's wife, Umm Talal, call it simply: the Bridge.'

It would be ten years before I saw the river from the other side.

The night before my departure for Jerusalem was an unsettled one. My son woke with a nightmare, then one of my daughters woke complaining of a cricked neck. By 5.30 a.m., everyone in the household was awake. I packed for the journey grateful for the distraction of cartoons. My flight to Paris wasn't to leave until late morning, but the lack of sleep had made me uneasy, and I was at the airport early. One reason for the trip was to give a lecture at a hospital in East Jerusalem, al-Makassed, at the invitation of the lead cardiologist there, Dr Izzedein Hussein. Another was to speak at the Augusta Victoria Hospital, on the invitation of its chief of nursing, Dina Nasser. In the departure lounge I flicked through my lecture notes restlessly, unable to focus.

It felt strange to fly over the Scottish Borders, so empty, and into Paris, so congested. The gathering point of so many journeys from northern Europe intended for the south. It was already dark as we approached Cyprus, its coastline a kinked thread of fairy lights. Then unmistakably I recognised the Levantine strip – the hooked peninsula of Haifa and a plumb line of highway floodlights all the way to Gaza.

On the flight I read a novel: *Austerlitz*, by W. G. Sebald. The narrator of the book recounts a series of conversations with a Jewish refugee, Jacques Austerlitz, who reached England from Prague by way of Kindertransport on the eve of war. Long passages of the book are devoted to detailing the appalling atrocities perpetrated on Jewish people through the Nazi death camps, in particular Theresienstadt / Terezín, to which Sebald devotes one of the most extraordinary sentences in literature, over two thousand words long, which in English translation spreads over seven pages. Of that sentence, one literary critic commented: 'Sebald's literary aim is to find a stylistic equivalent to the monstrosity of the Nazi language by composing a monstrous sentence. Monstrous both in terms of form and content, that is to say, in its hyperbolic length and the replication of inhuman Nazi terminology.'

This fictional Austerlitz had a history similar to that of many Jewish children forced by Nazi genocide to flee: he settled in Wales and, after many trials, grew up to become an architectural historian fascinated by the creations of military engineers. 'The largest fortifications will naturally attract the largest enemy forces,' he tells the narrator of the novel. 'The more you entrench yourself the more you must remain on the defensive, so that in the end you might find yourself in a place fortified in every possible way.' It was 2018; Israel then had the highest military spending per capita in the world. Only two countries had a greater concentration of soldiers: North Korea, and Eritrea.

The city of Tel Aviv has engulfed the old port of Joppa. I walked through airport corridors that corralled me into crossed walkways clad in monumental stone, more palatial than municipal, the pedestrians dwarfed and silenced. Along one immense wall had been fixed a satellite photograph of the Levant emphasising the geographical unity, the indivisibility, of God's chosen land. It's always striking to see the borderlessness of the earth when viewed from the sky.

The hosts who had invited me to Jerusalem to speak had given me what they called a 'comfort letter', to show border guards. Eyes like pinhole cameras scrutinised the document, then the guard tapped something into a computer. I was waved through to a too-bright forecourt, where a driver in a drooping leather jacket as black as a nun's habit was waiting for me. His name was Louis; crucifixes hung from the roof of his Ford Explorer, and he drove me at 120 kilometres an hour down highways lined with palm trees, past signs with place names familiar from Sunday-school stories. We passed tourist signs marked 'Martyrs' in English and Hebrew, but not in Arabic.

As we entered the city of Jerusalem we approached what appeared to be a vast illuminated gateway, a silver-spun harp of cables suspended from a lofty floodlit mast. It was a rail-and-passenger bridge – the Chords Bridge, designed by the Spanish architect Santiago Calatrava – a geometrical vision of precision mathematics, a structure that might have emerged from the pure realm of forms. Against the darkness of midnight it was a luminous spray of light, a fusion of fluted ivory and marble.

Scant water flows in the Judaean hills, and this, Jerusalem's most impressive bridge, is positioned not to provide a crossing over a river, but simply to pack more layers of traffic onto the already congested modern city of West Jerusalem. Its angled mast was the highest structure in the city. I'd seen similar Calatrava bridges in Manchester and in Dublin, as well as his magnificent transportation hubs: a train station in Italy's Reggio Emilia, and the grandiloquent, cathedralesque station under New York's Twin Tower memorial. It was a shock to see such a statement of modernity tower over a city so ancient.

The streets became narrower, no longer three-lane highways but medieval mule tracks buttressed and expanded beyond what they were ever intended to bear. After we'd negotiated a checkpoint busy with soldiers, Louis relaxed and for the first time switched on the radio. The taxi swerved round the Ottoman walls of the old city, overtaking a moped driver in flapping white robes. Horn beeps were exchanged between darting traffic at Gethsemane, as the road curled beneath the Dome of the Rock. It was well after midnight: beyond the Garden everything became dark and quiet and Louis began crossing himself very quickly, muttering prayers. We passed the Mount of Olives, and shortly afterwards pulled up at an iron gate – the enclosure of a Benedictine convent. I messaged E. to tell her I'd arrived, found a bunkroom, and slept.

To reach the city I'd passed by its newest bridge; the following day I sought out its oldest. Not a bridge as such, but an arched structure over water all the same – a channel thought to have been hollowed out and roofed over four thousand years earlier. The first stone bridges we know of were aqueducts – their traces have been uncovered in the north of what is now Iraq – and this channel gave an insight into not only the skill of ancient engineers, but the realities of defensive borders, and the perennial need of human beings for water and security.

Bronze Age Jerusalem was, it seems, a place prone to violent siege, and the city's most reliable spring, if left to flow unimpeded, would let water leak outside what are its naturally defensible walls. Around 1800 BCE, even before Moses had that vision that all the land between the Jordan and the sea would belong to his people, a trench was dug for channelling Jerusalem's Gihon spring towards a pool where its water could be conserved in times of siege. It was a man-made river, built by Canaanites; a thousand-foot score through the rock, entirely bridged over with slabs. That Bronze Age channel is now dry, but nearby is a tunnel hollowed out a thousand years later – around 800 BCE – which still replenishes a nearby cistern. Both tunnel and cistern are described in the Old Testament of the Bible as defensive measures to protect the city against siege by an Assyrian army. Later the spring was named for the Virgin Mary, and was believed to be replenished by her personal intervention.

There were crowds queuing to get down to see the spring. Ancient rock arches are nothing unusual in Jerusalem, and I ran my hands over the stones, trying to sense some difference in the arrangement of this almost three-thousand-year-old arch from those constructed by Bronze Age Hebrew refugees from Egypt, by imperialising Romans, by expansionist seventh-century Muslims, by crusading Europeans, by Abbasids, Mamelukes, Ottomans, Brits, Israelis – the successive peoples who have laid violent claim to Palestine. But all of them just felt like stone, cut and placed expertly by human hands to meet human needs: water, shelter, safety.

News filtered through from the catastrophe of Gaza, a walled-in strip of territory under siege for many years now. Bombs dropped by Israel had destroyed its sanitary facilities, and repair was impossible because machine parts were not being allowed into the city. As a consequence the sea off the coast of Gaza was in receipt of thousands of gallons of untreated sewage. Longshore drift knows no international borders; it takes a particular kind of bridge-thinking to realise that commons such as water and air are by necessity under the stewardship of everyone. A Qatari initiative was, I heard, being allowed in to arrange repair of Gaza's sewage plants, not for any humanitarian reason, but because faeces were starting to wash up on the beaches of Tel Aviv.

* * *

In Jerusalem the police position themselves in front of the old Ottoman fountains, which are sadly all dry. Those sites must have been selected centuries ago as nodal points of the city, for maximum convenience of its citizens. It seemed grimly perverse to situate police at those places where for centuries Jerusalemites have been obliged to pass and congregate, to gather life-giving water.

From the viewpoint at the top of the Augusta Victoria tower in East Jerusalem, it's possible to see three borders – that of Jordan with the occupied territories, that of the occupied territories with the city of Jerusalem, and the partition of the city itself. The new concrete wall being built by Israel doesn't follow the boundaries agreed in 1967, but loops back and forth in a haphazard line. I wondered whether its path had been chosen to better achieve military objectives, such as surveillance of the population and access to water, and asked one of my hosts. 'It is doubtful whether the wall achieves any military objectives,' he said. 'Certainly this was not what determined its route. The main consideration for the path taken by the wall is to keep the maximum Palestinian land within Israel with the minimum number of Palestinian inhabitants.'

Downstairs from the tower, on my way to the hospital's auditorium, I passed a framed print of Rembrandt's *The Anatomy Lesson of Dr Nicolaes Tulp*, a reproduction of which also hangs in the entranceway of my own medical school in Edinburgh. Its presence here in Jerusalem spoke to the universality of medical practice, and of the study of anatomy. Tulp revealed to his enraptured audience how the differences between human beings are illusory; anatomists know that race is a myth, something now confirmed by geneticists: there is nothing beneath the skin that identifies one people as different from any other. It's now known that there is more genetic diversity within the ethnic groups of sub-Saharan Africa than there is between all the ethnic groups of the rest of the world. W. G. Sebald wrote of Rembrandt's painting, 'The anatomy lessons given every year in the depth of winter by Dr Nicolaas Tulp were not only of the greatest interest to a student of medicine but constituted in addition a significant date in the agenda of a society that saw itself as emerging from the darkness into the light.' Tulp was Dutch, but his students were drawn from across Europe, pulled towards the city of Amsterdam for its spirit of open enquiry, for its belief that dissent in matters of religion shouldn't be allowed to distort politics.

The burden of the conflict hung like a pall over Jerusalem, making the prospects for peace feel dismal, even hopeless. The air was electric with fanaticism and distrust; fear and the frenzy of possession lay over the city like a delirium. I learned that, uniquely to Israel, citizenship and nationality are understood as distinct categories in order to privilege one group at the expense of others. The East Jerusalemites I met told me that the state considered them permanent residents of Israel, but without citizenship. 'Israel is not a state of all its citizens,' said Binyamin Netanyahu once of the distinction. 'Israel is the nation-state of the Jewish people and only it.'

When General Allenby defeated the Ottoman garrison of Jerusalem in 1917, he issued a proclamation to the people: 'The hereditary custodians at the gates of the Holy Sepulchre have been requested to take up their accustomed duties in remembrance of the magnanimous act of the Caliph Omar, who protected that church.' The keys to the church are still held by a Muslim family, in part because the Christian sects who worship there have always fought over privilege of access. Watching the people kneeling, wailing, crawling through this apogee of holiness I felt only that this literalism in the interpretation of religious texts seemed wholly at odds with their intention. It could even be thought of as absurd, if its consequences were not so tragic.

The roads were quiet when I took a taxi to Ramallah. 'Sabbath,' said my driver, 'Jew holiday,' then rolled his eyes, afraid to say more. 'Jew*ish* holiday,' I wanted to correct him, but in Israel or in the West Bank I was always afraid of saying the wrong thing. On my way through the high wall of reinforced concrete that divides the occupied territories from East Jerusalem I observed that there were separate, racialised entry and exit points for Israelis and Palestinians. Security cameras were propped on every vantage point, as if to immortalise and emphasise the fear generated by the wall.

At the checkpoint a teenager cursed with acne put on a show of boredom as he went through the contents of my bag. I felt a paternal impulse to offer him some dermatological advice; his own fear was palpable, and reminded me that fear and aggression occupy the same circuits in the brain, and have the same triggers. Psychologists have shown that the more fearful a human or animal is, the more aggressively they behave. Checkpoints too are bridges – conduits between lands that would be otherwise unconnected, *dis*connected. They are structures engineered to be swiftly closed, to keep people apart. A network of cylindrical watchtowers, each driven into the bedrock like a concrete nail, worked together with the wall as effectively as a gulf or estuary.

After an hour's delay I was through, to a city of rutted roads, stray dogs, burning trash, wheelless cars at the roadside. Ramallah seemed a place of contrasts – of power but also of weakness, of determination but equally of despair. To me it seemed frantic yet paralysed, as if caught in a nightmare from which it couldn't wake. Road signs and buildings were pockmarked with bullets, and I noticed the extent to which the roads of that part of the West Bank are doubled, as if everything in the landscape is viewed through a trick mirror. The old roads curve earthbound, keeping faith with the land's contours, showing their allegiance to the stone bridges and olive groves they serve. Above them soar new concrete highways on pillars, gliding without concern for the landscape, paying no heed to river or ravine. These are roads built for new settlements, and like the swifter entry and exit points in the wall are only for the use of the privileged group. Israeli citizens are offered low house prices and other incentives to move into these settlements, which are illegal under international law.

The settlement highways are bridges too, in a way; bridges between islands of habitation and fear. But they are at the same time the antithesis of bridges, in that only one group of inhabitants is permitted to use them.

* * *

As a doctor committed to the principle of medical neutrality, and interested in the worldwide fraternity of medicine, I felt an immediate sense of familiarity on arriving in the village clinics of the West Bank, north of Ramallah. In the waiting rooms there were public health posters; in the clinics there were thermometers, blood pressure cuffs, trolleys for taking blood, examination couches. Each had a small laboratory, the machines identical to the ones I remember from my own training. I helped out in a vaccination clinic, impressed by the assiduous record-keeping, the comprehensive vaccination regime that was aspired to – the universality of good-quality healthcare.

One doctor I met told me he'd been unable to study medicine anywhere in the Levant; instead he'd studied first in Yugoslavia, then in France, then in Norway, learning the language of each adopted country as he went. Another I met had studied in Belarus and spoke English, Arabic, French, Russian and Belarussian with ease. The borderlessness of his training was striking, as was the imposition of the borders under which he had to operate. Whereas my clinic in Edinburgh is funded through general taxation on one of the world's busiest economies, this clinic was supported by Japanese aid. The Japanese who had sponsored the King Hussein Bridge. It felt to me significant that such a distant government had chosen to support both bridges and healthcare as ways of bringing people, and keeping people, together.

One of the nurses took me out on her rounds. We visited a woman paralysed from birth, catheter-dependent, obliged to use a wheelchair, yet who lived in a house unreachable except by steps on the summit of a hill. She wore impeccable hijab, and told me in a tiny, high-pitched voice of her gratitude for the care she receives – the welcome with which good medical care is received almost everywhere. The needs of someone who has been paralysed are borderless – physiotherapy, catheter care, skin care, infection control. As we checked her legs for ulcers, I thought of patients in Edinburgh for whom I've done the same, of others across the borders in Israel and in Gaza for whom the same is done every day.

* * *

A view on the Jordan

The night before leaving Israel, I attended a poetry reading in Haifa, where a Palestinian poet explained that now seventy years had passed since the foundation of the state of Israel, he no longer wanted to hear about Nakba, the 'catastrophe' in which his grandparents had been forced from their land, by a people who had been forced from *their* land, and apartheid between groups of peoples established. He didn't want to hear any more about those thousands of people who'd had to flee for their lives over the Allenby Bridge into Jordan.

Khalas Nakba, he said, 'enough about the catastrophe'. The focus should, he added, be on finding a way to break down the division, accepting that the land between the sea and the River Jordan was now unequivocally controlled by one power, and to fight for the principle that all human beings on that land be honoured with equal rights. His priority was less that of finding a way to undo the occupation, than of undoing the assumption that one group of residents should be perpetually privileged over all the others. Until that happened, Palestinians would remain tangled in the snares that the rulers of the land had set for them – snares that are tightened every time they struggle, but are never slackened. As a European, I felt culpable, as if the peoples of the Jordan valley are being forced to pay in land, lives and freedom for atrocities committed not along the banks of their own river, but alongside the Rhine, the Danube, the Vistula, the Po.

Chapter Twenty-Two

Bridges of Empire

Neretva / Miljacka / Drina, Bosnia
Mostar Bridge: stone single-arch (1566, destroyed 1993, rebuilt 2004), 29m
Latin Bridge: stone 3-arch (1565), 24m
Mehmed Paša Sokolović Bridge: stone 11-arch (1579), 379m
Borders: Muslim–Catholic–Orthodox communities

In a war in which multi-ethnicity was itself the enemy, the destruction of the bridge appeared to mirror that of the multi-ethnic ideal of Bosnia – a place almost defined by bridge-building – between communities, between nationalities, between faiths.

LAURA SILBER & ALLAN LITTLE, *THE DEATH OF YUGOSLAVIA*

Geography is destiny: Bosnia is a land of flinty mountains cut by ravines, and for most of its history was undeveloped, seemingly undevelopable. To the Romans it was 'Illyria', a place largely spared the upheavals that rocked more easily invaded parts of Europe. From the 1300s that began to change: Turkish incursions from the east and a solidifying sense of Serbian nationhood made it by turns the bridge and the border between the Christian West and the Muslim East. To drive from the outskirts of many a Bosnian city towards its centre is to take a journey back in time. First comes modern, aid-supported glass-and-steel, a response to the war of the 1990s between Croat, Bosniak and Serb. Then comes a ring of Yugo-communist concrete architecture, a legacy of the Cold War years. Then Austro-Hungarian buildings from the days of Habsburg rule, proud, square and confident, claiming the Balkans for Mitteleuropa. And finally Turkish and medieval buildings – the smallest and most irregular, fitted to a time of horses, carts, and sheep in the city streets.

Evelyn Waugh's reported quip about the former Yugoslavia, 'see Dubrovnik and Split', is still funny, though its implication that only the coast of the country is worth bothering about less so. It's a coast now divided by borders that weren't there in Waugh's time: Slovenia, Croatia, Bosnia Herzegovina and Montenegro. The portion of coast allocated to Bosnia is tiny, just a few kilometres across, at the port city of Neum. That territory divides Croatia in two, but gives the new nation of Bosnia Herzegovina an umbilical connection to the water's edge.

The people of Neum are mostly ethnic Croats, and until Croatia joined the EU they behaved as though they lived in that country – using Croatian currency and working for the most part in Croatia. At the time I visited, 2019, a bridge was under construction to join a peninsula of Croatia's southern coastal strip to its northern mainland, circumventing the short section of Bosnian territory, and uniting the country by road. It was to be a 2,400m cable-stay bridge of six towers, and a bridge of contradictions: it marks a failure of cooperation between neighbours, but mitigates the absurdity of Balkan borders; it's funded by the EU, but built by the Chinese state. Though Croatia has some of the finest welders in the world (Croatians were hired to work on the new Queensferry Crossing in Scotland), I read that the workers for the Pelješac bridge were flown in from the Far East. Some Bosnian politicians opposed the bridge and refused to grant consent, insisting that it limits their only port's access to the sea. Croatian politicians claim the bridge is essential for their national development, adding that large ships bound for Bosnia unable to pass under it could simply use one of Croatia's many ports instead.

Again and again in the former Yugoslavia I was to come upon the conviction that bridges are not an unmitigated good. Ease of connection can bring a community more difficulties than it resolves.

* * *

The Dalmatian coastline appeared scorched, almost waterless. At the seafront of Split giant ferries stood in their wharves, their prows blunt and chunky, like darning needles waiting to stitch the mainland of Croatia to its belt of islands. Along the promenade would-be Instagram influencers tossed their hair unselfconsciously, pouting for tiny lenses.

The Croatian military was having an open day. You could get your picture taken shouldering a bazooka, trying on camouflage fatigues, or sitting at the controls of a chopper. During the Cold War, early 1980s, I remember the British army doing the same in my small Scottish town.

My companion for this journey would be Allan, a journalist who for years during its war lived in the former Yugoslavia – he would be my bridge to the country's past, and to some of the people who lived through that war.

On the way to the town of Mostar we stopped at Medjugorje, where in 1981 a group of six young people reported having experienced a vision of the Virgin Mary. As a result the village had become a boom town: Catholic pilgrims from all over Europe jostled between gleaming tour buses and the main square; both freshly built and refurbished churches had been cut with sparkling new stone. There were stalls selling holy water, crucifixes and illuminated icons of the Virgin. Pope Francis has repeatedly voiced scepticism and caution about Mary's visitations, but the tourists come nevertheless. Medjugorje is in a region of Bosnia dominated by ethnic Croats, and in 2021 the president of Croatia discussed the issue after a meeting with the Pope: 'It would be good if Our Lady would stop appearing every day,' he said. On the street I passed an Italian woman wearing sunglasses and red high heels. 'This place is awful,' she was shouting into her phone, 'it's thick with nothing.'

As we walked back to the car and prepared to drive on, Allan told me about a visit he had made to the town during the war. Massacres of Bosnian Muslims were ongoing to the east, and outside the main church he'd seen a long queue of soldiers, each waiting in line to receive absolution from the priests inside.

* * *

The English word 'bridge' comes from an old Indo-European root meaning 'log' or 'beam'. In many Slavic languages '*most*' comes from the same etymological root – it means 'bridge', but there are echoes of that derivation from trees, or logs, in the English word 'mast' (i.e. the mast of a ship). The old Ottoman bridge at Mostar took nine years to build, starting in the late 1550s; on 9 November 1993, four years to the day after the fall of the Berlin Wall, it was destroyed.* The River Neretva, which flows 24 metres below it, had become the front line in a battle between Croatian and Bosnian fighters.

Allan told me about a young Croat soldier he met in 1992, who said of the Neretva: 'Listen, on this side of the river we are in Central Europe. We are in the civilisation of Mozart and Goethe. Over there, on the other bank, is the East, Bosnia, Serbia, all the way to Saddam Hussein and that Asiatic way of thinking.' For that soldier, the Neretva was a divide as freighted as the Hellespont – a boundary between worlds.

At the close of the war an alliance of the World Bank, UNESCO and several European governments donated funds for the reconstruction of Mostar's old bridge – it was too important a symbol to be left in pieces. A Turkish engineering company specialising in the repair of Ottoman bridges was commissioned; the shattered stones were pulled from the opalescent torque of the river. They had been so badly damaged that few could be reused.

The bridge at Mostar was built by the Ottoman Empire as part of its consolidation of power, symbolic of its hold over the Balkans, so from the perspective of some Christians in the area (whose trade and cultural connections were to the north with Austria, and west across the Adriatic with Venice), it was not a good bridge – it was a bridge that helped bring in Muslim armies from the east, and facilitate tax collectors. Perhaps that's another reason why the Croat army blew it up. The Albanian novelist

---

* The commander who ordered the destruction of Mostar's bridge, Slobodan Praljak, had been a theatre director before the war. In 2017, he stood in the dock in the Hague, took out a vial of poison and drank it live on the courtroom's CCTV in a final *coup de théâtre*.

Ismail Kadare captured that Balkan distrust of bridges in his novel *The Three-Arched Bridge*, which describes the arrival of the Turkish Empire in Albania. The bridge is a 'stone collar' on their river, the 'devil's backbone', he wrote; 'in fact, all great building works resemble crimes, and vice versa . . . there is no difference between the two'.

The bridge at Mostar wasn't much of a military asset – some Croats claimed that the Bosniaks were using it to move weapons, but its real offence was as a reminder of the many Ottoman legacies that endured on land they regarded as Christian and Croatian. The bridge has been framed as a symbol of contact, dialogue and harmony between peoples, but also as a symbol of the reach of the Turkish Empire.

The approaches to Mostar's bridge are like Escher stairways, passing beneath towers once occupied by the *mostari*, bridge guards, but where an

exhibition of photojournalism now memorialised some of the many atrocities of that war. It was a surprise to see this supposed symbol of harmony between peoples used as a gallery of conflict. I stood at the crest of the bridge's arc; its stone was pale and lustrous, as if sculpted from the rock of a full moon, both taking and emanating light.

Legends of human beings incorporated into the stonework as sacrifice are common among the Ottoman bridges of the Balkans, and there's one at Mostar too: that during its construction two lovers fell in and were immured within its spandrels – one to the western side and one to the east. The story told that the longing between the two of them held the stones of the bridge in tension. It's been estimated that before the war, a full third of marriages in the town were mixed across faith lines. It is possible to focus on that third as a positive sign of community integration, but it's equally possible to focus on the fact that after fifty years of communist rule, two-thirds of the population preferred to marry only among their own.

To stand on the arch of Mostar's bridge was to be jostled by tourists from Italy, Bosnia, Serbia, Croatia, China – all like me in awe of the grace and the span of its elegant arch, at how high it stands above the water, at the visionary execution of will that it be repaired. A man in a thin wetsuit sat astride the parapet, ready to leap off for a few euros.

On a hill high above the town was propped a monumental cross – an unequivocal statement of how for many of the local people the Neretva is a Christian river. Allan told me that the two communities are still largely divided by the water – Christian to the west and Muslim to the east. Some of the tourists were enjoying the sun, the view, their ice cream. Others looked thoughtful, even mournful; the connections between stones being so much easier to repair than those between people.

The suffix '*evo*' means 'place of', and '*saraj*' is from the same Turkish root as '*caravanserai*'. Sarajevo means 'the stopping place of caravans'. It was said of the town that it was the furthest into Europe a camel train could reach before the animals became sickened by the climate and were obliged to return south and east. Sarajevo was where their loads would be transferred to horses and pack ponies for the journey on into Europe. The city's foundation was as a bridging place between East and West.

The Miljacka River flows through the centre of the town, and over it is a short, functional bridge where Franz Ferdinand, heir to the Austro-Hungarian Empire, was assassinated in 1914 by Gavrilo Princip. The bridge is an old Ottoman one, now known by its original name: the Latin Bridge – Latinska Ćuprija – because it helped the Catholic population access the heart of the city (as '*most*' is the Slavic word for bridge, '*ćuprija*' is Turkic). The Latin Bridge marked the boundary line between Christian and Muslim communities. But between 1945 and the war it was known as Principov Most, to celebrate the name of the assassin who liberated the country from Austria-Hungary. 'The city council brought back the old name of Latin Bridge during the war,' Allan told me, 'when it no longer seemed appropriate to celebrate a Serb fanatic with a gun. There were real live contemporary Serb fanatics raining hellfire down on the city from the hills above.' Ferdinand was assassinated at the point where the bridge meets

the northern, Muslim bank of the city, and as I walked across it to the south I wondered at the way these few piers and cobblestones could be the hinge upon which Europe's twentieth century had turned.

Beyond the river I continued on south and up into the hills overlooking the city, towards the high ground once occupied by Serb snipers. The call to prayer began to echo between the walls of the valley; I noticed 'Remember Srebrenica' daubed on a wall mural, and posters condemning the indifference of the EU. Many Bosnians, it was clear, feel they have been abandoned by the West.

In the necropolis of Sarajevo a majority of the graves were marked 1992, 1993, 1994, 1995, 1996. The intense need for grave space during those years turned a local football pitch into a cemetery. Before a diminutive shrine to Princip, Allan told me that in his experience of war, everyone's first priority is to secure the safety of their children. Only once the children are safe do other loyalties begin to creep in – more distant relatives, colleagues, neighbours, community. The last priority of all is nation.

The Bosnian population is estimated to be 50 per cent Muslim, 30 per cent Serb Orthodox, 15 per cent Croatian Catholic, and the constitution divides representation between the groups. Two Sarajevans – one Roma, one Jewish – protested that division at the European Court of Human Rights in Strasbourg. In my conversations with Bosnians, and in the slogans sprayed on the walls of the city, a raw sense of grievance persisted, barely suppressed fury shimmering between the renamed street signs. The old mosque in the city centre had been restored by money from the Gulf, its stonework as glittering, new and redolent with investment capital as the churches I'd seen in Medjugorje.

The rivers were running red with soil after the rains – tunnels, gorges, ravines, forest. There were many inscriptions on the concrete bridges commemorating their construction in the communist years. The road was hypnotic, and highway markers ticking past as Allan and I drove east towards Serbia. As we approached Višegrad the tributaries of the Drina had begun to lighten in colour. Though still opaque, the water gradually took on the colours of the sky until by the time we reached the city the river was a cool chalky blue. It flowed north, muscular and glistening, twisting through the tight valleys. Višegrad is the last town before the Serbian frontier, in that autonomous enclave of Bosnia in which the Serb population is strong, the 'Republika Srpska'. Just a few miles beyond it, the river would *become* the frontier, and stay that way all the way to the River Sava.

The English language has many synonyms for 'walk': pace, peregrinate, strut, swagger, shamble, shuffle, toddle, amble, saunter, stroll, stomp, tramp, trudge, prance. But for all the richness and texture of the language, it lacks a word specific for the surface on which one moves. I've often wished that, like certain Native American languages, there was a way of communicating the special nature of a walk over different terrains. It would be splendid to have a word for the sensation of walking over a bridge, for the way bringing air and water beneath your feet changes your relationship to space. This new word would communicate something of that feeling alluded to by Ivo Andrić in the novel about Višegrad for which he won the Nobel Prize, *The Bridge Over the Drina*: 'Even the least of the townsmen felt as if his powers were suddenly multiplied, as if some wonderful, superhuman exploit was brought within the measure of his powers and within the limits of everyday life, as if besides the well-known elements of earth, water and sky, one more were open to him.'

Višegrad was sun, bridge, white stone, blue water, and dark hills of velvety green. I walked out over the bridge made world-famous by Andrić. Kids begged from tourists; one aged eleven or twelve held another, aged just four or five, over the stone parapet as if to drop him into the water, cackling with laughter as his victim squealed in distress. Bridges are public platforms that can seem to encourage dramatic performances, and the bridge over the Drina has often been chosen as a setting for the grim theatre of war. During the war it was reported that Muslims were thrown off the Višegrad bridge by Serbs, used as target practice as the current pulled them away towards the frontier. Some of the murders of Muslims were framed as retribution for vandalism of a statue of Andrić, who, though a professional historian and diplomat, has been criticised for perceived anti-Turkish and Islamophobic sentiments in his writings.

Cloud obscured the sun for a moment, and it was as if the water blinked. I took my phone out to take a snapshot and the kid pulled his victim back over the parapet, throwing him aside to follow me instead, muttering, 'iPhone, iPhone, iPhone' in a maddening sing-song voice. The war had ended more than twenty years before, but Višegrad still felt like a place where poverty and desperation lay close beneath the surface.

The first pages of Andrić's novel tell the story of Sokolović Mehmed Pasha, who commissioned the bridge in the sixteenth century. He was a Bosnian conscript taken from a very poor village in the area as tribute, to stand service to the sultan. He rose up the Ottoman ladder of power to become a vizier – as high as it was possible to go – and one of the wealthiest men of the Turkish Empire. Andrić imagines him as a child waiting in misery for a boat to take him across the Drina into captivity, dreaming that one day there might stand there instead a glorious bridge of stone. Without ever returning to his homeland, in his old age Sokolović commissioned such a bridge, transforming the pain of division from his family into an enduring symbol of connection. It's not certain if he ever saw the bridge that would become his legacy, or even how closely he was involved in supervising its construction. In Andrić's novel, the bridge's engineers are Italians, its foreman a sadistic, bloodthirsty Turk; with gentle patience across three hundred pages it tells the history of Bosnia through the vicissitudes of the bridge.

Allan and I ate in a hotel overlooking the water, the latest iteration of the inn described by Andrić, which was itself preceded by a stone caravanserai built at the same time as the bridge. The restaurant was empty but for a huge party of Chinese tourists who had ordered a banquet of three times more food than they could eat. Serbia is one of the few countries in the world it's possible for Chinese citizens to visit without a visa, and the frontier between Serbia and Bosnia's Serbian entity is porous.

I joined them later outside, taking selfies with the local celebrity – the bridge now almost half a millennium old. It was a bridge of contradictions: built by an enslaved Christian who made good in the Turkish capital, a symbol of progress but also of exploitation, of the uneasy border between

Serbia and Bosnia's autonomous Serbian enclave, of the indelible connections between West and East, between Muslim and Christian worlds; a reminder of the horrors of war and the possibilities of peace. 'Thus the generations renewed themselves beside the bridge and the bridge shook from itself, like dust, all the traces which transient human events had left on it and remained, when all was over, unchanged and unchangeable.'

The wall of mountains on the Serbian side of the border was black under a geology of rain clouds, dark as earth. On the east bank of the Drina was the pop-up theme park of Andrićgrad, like a toy-town film set for a movie of Serb supremacy. Behind a glorious new Orthodox church, funded by a Serb millionaire, you could buy fridge magnets of Princip, transformed into a hero of Serbian, not Yugoslav, nationalism. Before the war, the town had been predominantly Muslim; now it was almost exclusively Orthodox. Sprayed along one wall of Andrićgrad was an immense mural of Princip beside a text that read, 'Life is an inexplicable miracle, because it is spent away, yet it lasts and holds as still as the bridge on the Drina.'

Medjugorje, Sarajevo, Višegrad: the new religious buildings of each place were in their way testaments, and statements of belonging. Perhaps they were also statements of intent. It did not feel as if the bridges of Bosnia brought its people closer to peace or reconciliation.

2020s

Chapter Twenty-Three

Bridge of Reconciliation

Foyle, Northern Ireland
The Peace Bridge: cable-stay (2011), 235m
Border: Catholic–Protestant districts of Derry / Londonderry

The EU threatens no one's freedom or independence, it is the way in which we collectively advance the cause of freedom, prosperity and peace.

JOHN HUME, 'EUROPE AS A FORCE FOR CREATIVE RECONCILIATION'

In the pandemic summer of 2020, after months of grounded aeroplanes, E. left the country to visit her relatives in Italy for the first time in a year. I borrowed a camper van and drove with my three children to Stranraer in the southwest corner of Scotland, then sailed across the North Channel to the port of Larne. Over the previous ten years I had learned that no amount of solo travelling could prepare me for the challenges (and pleasures) of travelling with children. I learned to curb my ambition as to how far we could comfortably manage in a day, never plan to do more than two activities between morning and evening, and always, always pack a bagful of snacks.

We were greeted in Northern Ireland by a traffic island in the shape of a gigantic crown, and rows of Union flags. Interspersed among all the red, white and blue was a single Israeli flag with its six-pointed star. It was baffling to see a flag of the Levant drafted in to the island of Ireland as a surrogate marker of belonging.

We drove up the Antrim coast with Scotland visible on the northeast horizon – the tip of the Kintyre peninsula. This would have been the Irish exit and approach for the bridge proposed by Boris Johnson, then the UK's prime minister, a modern Giant's Causeway across a channel two hundred metres deep. When asked about prospects for a bridge between Kintyre and Northern Ireland, Scotland's then First Minister replied: 'If he's got £20 billion to build such a bridge going spare at the moment, that could be spent on more important priorities.'

Trees stood in groups by the beach, leaning towards one another like politicians at a summit; the towns we drove through were enjoying peak holiday season. At the Giant's Causeway we hopped and clambered over sixty-million-year-old columns of basalt, similar to ones I'd seen across the strait in Scotland. My memory of the old legend was patchy, but my daughter recounted it for my benefit: how the Irish giant Finn MacCool built a causeway to fight a giant in Scotland, but on seeing the size of him took fright. His wife back in Ireland had a brainwave: Finn should dress as a baby and get tucked up in a giant cradle. The Scots giant stormed down the causeway into Ireland, glimpsed the size of this baby and took fright himself at the thought of how big the father must be. He fled home, tearing the causeway apart as he went.

It's a story about Irish canniness and Scottish gullibility, but also about Irish vulnerability in the face of repeated invasions and humiliations from the east. MacCool had no intention of rebuilding the causeway after his near-miss. Some bridges, the myth seems to say, bring only destruction and exploitation, and are better left in ruins.

We were exploring the Giant's Causeway when a run of fine weather began to break. A drenching fall of rain was forecast as we packed up the camper van and drove to the River Foyle. There was a bridge over the river I wanted to see, and though I'd been in Belfast many times, I had never seen the city of Derry – or Londonderry, depending on your political allegiance. A year earlier E. and I had been devoted followers of a TV sitcom about the town, *Derry Girls*, which managed to wring laughs from the lunatic situations forced on the people of this city by the savage slow-burn civil conflict in Northern Ireland, euphemistically known as the Troubles. The rain was forecast to be relentless, and I wished I'd brought waterproof trousers for me and the kids, then started chuckling to myself, remembering a particular episode of *Derry Girls*. It's an exchange between Erin, a teenage girl, and her mother as Erin packs for a weekend away where Catholics and Protestants will meet to work together on a peace-building project.

MOTHER: Oh my God, don't forget your waterproof trousers! They nearly bloody bankrupted us! Do the Protestants have to bring waterproof trousers or will the Catholics be expected to do all the dirty work?

ERIN: It's not dirty work – it's an outdoor pursuits weekend.

MOTHER: I thought you said you were building bridges.

ERIN: Not real bridges, Mammy, metaphorical bridges.

MOTHER: Then why can't you wear metaphorical trousers?

In Derry a kindly traffic warden with a Scots accent let me park my camper van overnight in the city centre. On the radio I heard that John Hume, the Nobel Peace Prize laureate, politician and civil rights activist, had just died. His funeral would be held in Derry Cathedral the following day, and because of the pandemic restrictions the gathering would be limited to a few close family, friends and statesmen – all come to commemorate the passing of the only person ever to have been awarded the Nobel Peace Prize, the Gandhi Peace Prize and the Martin Luther King Peace Prize. In the streets of the Bogside, on the way up towards the cathedral, my kids were silent, wide-eyed at its murals of hunger-strikers and gunmen. A house-sized sign that read 'Welcome to Free Derry' had been adorned with an NHS logo, in acknowledgement of the pandemic. Over its blue heart a photo of John Hume had been taped, alongside an 'RIP'.

The day of the funeral the kids and I walked a loop of Derry's walls, sheltering under our umbrellas, looking down on the rain-black roofs of the Catholic area of the Bogside and the beetle-black cars of the funeral cortège. We paused at a memorial plaque about the siege of Derry and the 'apprentice boys' – an event described by Susan McKay, the author of *Northern Protestants: On Shifting Ground*:

> In the winter of 1688 the Catholic forces of the recently deposed James II surrounded the largely Protestant city. Its governor, Robert Lundy, wanted to negotiate surrender as they didn't have the resources to withstand a prolonged siege. But thirteen apprentice boys defied him and closed the gates. Lundy was banished, replaced by a stauncher man. The siege lasted 105 days. Derry's inhabitants were reduced to eating dogs 'fattened on the flesh of the slain Irish', horses, rats and tallow. Fever swept through the city. Thousands died. But there was no surrender, and the city was finally relieved by the new king, William of Orange. Every year, the Orange Order, the Apprentice Boys and others celebrate the victory by burning an effigy of Lundy to shouts of 'no surrender!'

For some people in Northern Ireland, peace has become synonymous with surrender – one of many reasons that the conflict has proven so enduring. In a book by Lewis Hyde called *A Primer for Forgetting*, I learned how some communities have managed to choose a different path, finding a way to live with ancient hurts and breaking cycles of violent retribution. Hyde also writes about the work of the Turkish Cypriot psychiatrist Vamik Volkan, who has developed the concept of 'chosen trauma' – an event that binds a community's sense of itself together even as it can close off the possibility of healing and making peace. Volkan has worked professionally with many of the conflicted groups I encountered on the travels recounted in this book – with Arabs and Israelis, with Serbs and Bosniaks, with

Turks and Greeks. The questions he asks of any group who have become locked in conflict are deceptively simple, but to answer them is to open up new possibilities of reconciliation, even in the most stubborn conflicts: 'How can the symbols of chosen traumas be made dormant so that they no longer inflame?'; 'How can group members "adaptively mourn" so that their losses no longer give rise to anger, humiliation and a desire for revenge?'; 'How can a preoccupation with minor differences between neighbours become playful?'; 'How can major differences be accepted without being contaminated with racism?'

Hyde's book also taught me about the work of the Irish peace activist Edna Longley, who said memorably of her fellow citizens, 'We should erect a statue to Amnesia and forget where we put it.'

Covid restrictions meant we couldn't go near the cathedral, but later, in a transcript of John Hume's funeral, I read the messages of condolence that had come in from the Dalai Lama, the Pope, Bill Clinton, and the homily from Father Paul Warren: 'In a time in our world when often small-mindedness and self-focus seems to be the driver, John never put anybody or any specific group first. He put everybody first. He didn't focus on difference and division. He focused on unity and peace and giving that dignity to every person.'

I walked with the kids out onto the Peace Bridge. Its cables hung like tripwires for Finn MacCool or his Scottish adversary; staggered towers leant towards one another as if reaching for an embrace. Its decks are slender, for pedestrians and bikes, curved in a sinusoidal wave between the Protestant area of Waterside and the Catholic Bogside. Its £14 million cost was paid for by the European Union's Special Programmes PEACE III fund – a European Union that 78 per cent of local people had voted to stay in, but from which earlier that year they had been forcibly ejected, thanks to the deciding votes of a handful of Unionist Northern Irish politicians.

The river flowed north beneath our feet into Lough Foyle; at its estuary five miles downstream it became the border between the UK and Ireland, and now the border between the UK and the European Union. The rains

had brought a windfall of branches, leaf litter, *litter* litter, refuse and detritus all passing with thousands of gallons of water under the bridge.

With the implementation of Brexit, the Good Friday Agreement had been all but ripped up. The architects of Britain's exit from the European Union have so far shown little appetite to explore alternative ways towards peace between the different communities of Northern Ireland. Within two years of my visit to the Peace Bridge the Irish nationalist party, Sinn Féin, would become the majority party in Northern Ireland.

As we crossed that bridge in the summer of 2020, Father Warren was speaking about the life of John Hume: 'Even in the darkest moments, when people would have been forgiven for having no hope, John made peace visible for others. His vision revealed what could be and with time, and determination, and single-mindedness, and, yes, with absolute stubbornness, he convinced others that peace could be a reality.'

Chapter Twenty-Four

Bridges of Innovation

Rhine tributaries, Switzerland
Tavanasa Bridge: reinforced concrete single-arch (1904), 85m
Salginatobel Bridge: reinforced concrete single-arch (1930), 133m
Borders: Switzerland–Liechtenstein–Austria; EU–non EU

Reinforced concrete does not grow like wood, it is not rolled like steel, and has no joints like masonry.

ROBERT MAILLART, SWISS CIVIL ENGINEER

The Romans may have known about concrete, but they didn't reinforce it with metal. In the nineteenth century, concrete began to be stiffened by heating steel bars as they were laid into the setting cement – bars that, as they cooled, pulled the concrete into a tense embrace that conferred immense strength. Reinforced concrete was patented in 1867, but it wasn't until 1904 that its more extravagant possibilities were realised: that year the first skyscraper made of the new material was completed – the Ingalls Building in Cincinnati – and the Swiss engineer Robert Maillart used it to bridge the Rhine at Tavanasa in Switzerland. Three years earlier he'd completed a reinforced concrete bridge at Zuoz over the River Inn, but it was with the Tavanasa Bridge that he began to show just what this new material could do.

The Swiss have effected an economic miracle: with the help of their banks, chemical laboratories and civil engineers, they've turned a nation of dairy herders and clock makers into one of the world's richest states.

Maillart was one of the wizards of this transformation – he had little formal training but had an intuitive feeling for the way forces flow through a structure. He figured out a way to incorporate the arch of a bridge into its roadway, an innovation that made his bridges cheaper than any of his competitors', as well as stronger. The aesthetic of the age was that bridges had to be clad in stone, but Maillart insisted that his concrete remain unadorned. Because the arch and deck supported one another the former could be as much as two-thirds thinner than a conventional stone-arched bridge.

But Maillart's early bridges weren't well loved. Though the bare concrete shallow arch of his Tavanasa Bridge would be admired today for the strength and subtlety of its design, Swiss bureaucrats with their formative years in mid-nineteenth-century Roman aesthetics couldn't adjust their expectations. They wanted stone, and so for almost twenty years after the completion of his bridge over the Rhine, Maillart was obliged to focus instead on buildings, and water infrastructure, because he couldn't get the commissions for the many bridge designs he submitted. When in the 1920s the Tavanasa Bridge was wrecked in a landslide it was replaced by a more traditional deep-arched design in stone. Few locals mourned the innovative original.

The year after Britain finally exited from the European Union and Joe Biden replaced the border-wall-obsessed Donald Trump as president of the United States of America, I drove from Italy across Switzerland from south to north. It was the school holidays, and I was driving home with E. and the kids. On our way to Switzerland we had passed Genoa, where in 2018 a reinforced concrete bridge collapsed under a lightning storm with heavy rain, killing 43 people – a terrifying reminder that even bridges of reinforced concrete have to be maintained and supported with care.

The British diplomats and bureaucrats in Brussels had been unable to implement a logical impossibility: leave a club but continue to receive all the benefits of membership. Switzerland, a country of under nine million, manages the dance of partial benefit without membership by signing up to a series of obligations – a compromise seemingly impossible for the British politicians charged with implementing such an equivocal referendum result.

Leaving Lugano we began to climb north through forested Alpine valleys, over several of the kind of sinuous concrete bridges that Maillart pioneered. Worldwide these are known as 'Maillart bridges' in his honour. Rainstorms had turned the highway into a wash, and the mountainsides around us were necklaced in torrents of foaming falling water. Swiss engineering is still world-class: I could see why the Bhutanese turned to Switzerland when they wanted to rebuild the old bridge at Thimphu. Through driving rain we skirted the Italian border as far as Piz Corbet, then turned east towards the Lenzerhorn. At the service station near Maienfeld there were *Heidi* memorabilia on display, beside cuckoo clocks and slabs of chocolate. E. was asleep in the back of the car. 'I just have to make a quick detour,' I said to the kids, buried in their devices. 'There's a bridge down here that I have to see.' A murmur of assent, and I turned off the highway not in the direction of Tavanasa but towards Schiers, a few miles to its east, where an echo of Maillart's revolutionary bridge of 1904 survives.

The Salginatobel Bridge spans a gorge deep in the woodland that cloaks the Austrian frontier. In its design, Maillart built on what he'd learned in Zuoz and at Tavanasa to create something of extraordinary beauty and strength that was nevertheless the cheapest of the nineteen designs that were submitted to its commissioners. It was completed in 1930, and sixty years later became the first bridge built in concrete to be recognised as of international historic significance.

The kids were silent as I drove over the bridge, parked up, and jogged down a forest track to find a way to appreciate it from below. Signs warned me against climbing in from the abutments; I looked along a tunnel of braced concrete struts that reminded me of the passageways into the Great Pyramid. Plaques announced it as a World Heritage Site, one of only thirty engineering wonders of the world – a list that included the Forth Rail Bridge back home.

Maillart's bridge is magnificent – it has stood for almost a century yet looks timeless, simply fulfilling the demands of beautiful, functional design.

Its compromises with form and function are so subtly resolved that it seems in its grace to have made no compromises at all. I wondered what is so special about the Swiss that they can take the best from their connections with their European neighbours, yet with energy, creativity and innovation balance solutions and conceptions that are uniquely their own.

Chapter Twenty-Five

Bridges of Cooperation

Kattegat, Scandinavia
Little Belt Bridge: suspension (1970), 1,700m
Border: Jutland–Funen
Great Belt Bridge: suspension (1998), 18,000m
Border: Funen–Zealand
Øresund Bridge: cable-stay (2000), 7,849m
Border: Denmark–Sweden
Svinesundsbron: concrete single-arch (2005), 700m
Border: Sweden–Norway
The Infinite Bridge: timber and steel beam (2015), 60m
Border: land–sea

The Idea: a circle on the coastline instigates a movement, that changes the perspective on the surroundings, creating a sense of community and a place for contemplation. There is no beginning nor end only the horizon, the movement, and the space that is created between people.

THE INFINITE BRIDGE, ÅRHUS

BMWs, Audis and Mercedes chased us down the autobahn as if trying to outrun environmental legislation, or a zombie apocalypse. There was a car upside down on the siding somewhere between Bremen and Hamburg, but no one stopped to check on it. Beyond the infernal tunnels and perennial roadworks of Hamburg the landscape began to broaden, the road pointing due north, over the high bridges of the Kiel Canal and into Scandinavia.

The frontier with Denmark was deserted, and I pulled over into a lay-by to marvel at the lack of border infrastructure – a gift of the European Union. Beyond it, my son looked up from his book to notice that the verges took on a shaggy, unmown look, and the pylons became shorter and more elegant. It was our first proper holiday together without his mother and sisters, though the year before we had taken a motorbike trip from Edinburgh to the northern isles of Shetland over a long weekend. On the coast of the island of Yell we'd gazed east and begun to imagine a road trip to Scandinavia, and a thousand-mile loop of the Kattegat – the waterway that connects Norway, Sweden and Denmark.

The past few years had seen the steady dwindling of passenger ferries between the UK and Scandinavia, driven out by lack of trade, or support for tourism, or government subsidy, or all three. Old maps show ferry routes leaving from Aberdeen's harbour into various ports of Scandinavia; as a student in the 1990s I once took a ferry from Newcastle to Bergen. But that's all gone now – even Harwich to Esbjerg has been cancelled – as if those connections are no longer valued, and not worth supporting. England has turned her smooth, hard back on Denmark and Norway, while her fissured, softer Celtic coast looks with longing to the west, hoping for a closer relationship with America.

As there wasn't a single crossing open to tourists we'd been obliged instead to sail from England to the Netherlands and take a long detour through Germany. At home, with maps spread out across the living-room floor, the drive from Amsterdam to the ancient Viking capital of Roskilde looked as if it could be managed in a day – something confirmed by my navigation app. But at the Little Belt Bridge between Jutland and the island of Funen, just short of Odense, I began to feel dangerously tired.

Denmark's islands stand like a series of immense stepping-stones between Europe and the Scandinavian peninsula. There was a campsite on the shore of the next strait along, the Storebælt, or Great Belt, which separates Funen from the island of Zealand, and we turned in. On a beach sword-sharp with glossy flints we called E. 'Is that birdsong?' she asked from five hundred miles away, though the route we'd driven was more than

double that. 'I can hear Danish blackbirds singing.' The Great Belt Bridge, almost twenty kilometres long, was beautiful in the gloaming, lit up to our south like promenade illuminations. We popped up the roof tent and slept.

The travelling jamboree of the Tour de France would cross the flatlands of Zealand that morning, the second day of its race. We watched the riders flow in a river of Lycra, Roskilde to Nyborg, across the Great Belt Bridge. It is a crossing of two bridges and a tunnel, built over an island once used as a prison for women accused of promiscuity, back when Scandinavians were both more pious and more censorious. The labourers brought in to work on the bridge, spinning its cables at heights exceeding two hundred metres, were recruited from ski-lift companies in France. In ordinary circumstances bicycles aren't allowed over the Great Belt Bridge, and I envied the torrent of the Tour de France cyclists as they passed: their speed under the open freedom of the sky.

Roskilde was hectic, with crowds of supporters in yellow jerseys pushing bikes, some wearing padded hats with Viking horns – a fusion of France and Denmark, ancient and modern; two different but complementary visions of Europe. We hired bicycles and joined them, pedalling around a city famous for its Viking boat museum. There are relics there of ships that knew not just the Kattegat, but the Arctic straits of Greenland. I had been in Roskilde almost twenty years earlier, on a stopover between Iceland and Greenland. It was part of a journey I made around the European Arctic, the account of which became my first book, *True North*. My son humoured me in my enthusiasm as I told him about my last time in the city, twenty years being for him unfathomable lifetimes ago.

The Danes seem to enjoy falling into character as their medieval avatars, and a few miles outside the city I found Haralds Bro, a replica of a tenth-century bridge of oak, reconstructed by volunteers and schoolchildren over a meadow now high with summer weeds, busy with swallows. There were no other visitors. My feet on the timber deck startled a crane from the rushes; it flew off with a few archangel beats of its wings in the direction of Sweden. My son stayed in the car reading – his book more real to him than a lost empire of Danes – but I walked back and forth over the reconstructed bridge thinking of an age when boats shuttled non-stop between Denmark and England, harrying, threatening and warmongering, but also trading, connecting and supporting.

Later, at Helsingør, Hamlet's Elsinore, we swam within sight of Sweden, dodging Baltic jellyfish and the ferries that sail on an infinite loop between the two nations.

The most striking exhibits of the National Museum in Copenhagen emphasise the pull of earth against the transcendence of sky. There were bog bodies on display there that had been dragged from millennia of sleep beneath the peat, and in an adjacent room, an ancient bronze model of a sun-god chariot harnessed to a stallion of heaven.

During an attempt to buy relatively cheap Danish diesel, my credit card was rejected: mobile phone masts and fibre-optic cables between Copenhagen and London buzzed with my indignation as I tried to find a call-centre worker able to unblock my card. Only then could we approach the wonder-bridge to Sweden. The highway sank into a trench of concrete, rigid carriageways and railways slipped first under the earth, then under the Øresund – the strait between Denmark and Sweden. The sediments are too soft to support the weight of the water above, so the segments of the tunnel were prefabricated, sunk, then covered over with dredged shingle and armoured stone ballast. The carriageways and railways emerge after a few kilometres onto a man-made island, before rising onto girdered beams, then across a cable-stay bridge with rail tracks laid out on a deck beneath the road. A suspension bridge was ruled out as too flexible for the heavy railway freight the crossing is obliged to bear, and so the decks are buttressed in cable-stays instead. This bridge between nations has almost twenty thousand users a day – many of them Danes who have moved to Swedish Malmö, having been priced out of their home country by Copenhagen's booming economy.

From the shingle of Peberholm we drove slowly skyward, my son videoing with his phone at the windscreen, and giving commentary. The sea was blue and clear beneath us as we climbed; marine windmills to the south fluttered like blown dandelion clocks. The Øresund link bridge is unique for the environmental concerns that governed its construction; its elements were prefabricated in Malmö, but also in Cadiz, and it is half owned by each of the countries it joins. It's a model of cooperative bridge-building that I'd seen too rarely in my journeys around the world – one that insists on a balance of benefits for the peoples of each side.

On our descent from the bridge's crown we saw Santiago Calatrava's great tower block of Malmö, the 'Turning Torso', which stands on Swedish soil but seems to be twisting around in order to gaze on Denmark. We approached the pay station, where a screen was taped up, forcing us to reverse and try again in a new lane. A woman with a round face and blue eyelashes took my toll of 65 euros. 'What happened to the booth next door?' I asked her.

'People break them,' she said.

'What, in anger?' I asked, unable to imagine a Dane or a Swede in the throes of road rage.

'No,' she smiled, also finding the idea laughable. 'With their wing mirrors.'

We parked up for the night in woodland near a lighthouse that stood high on a western promontory over the Kattegat. There were many Arab families doing the same; their pitched tents were arranged around campfires along the fringes of the forest. Zealand was just visible on the horizon; light fell in patches onto the sea as if onto a spotlit dance floor.

The Royal Republic of Ladonia was named for the dragon Ladon, who in myth guarded the golden apples of the Hesperides. It has no visible border; just a square kilometre of seashore woodland set out by the artist Lars Vilks in response to the Swedish government's insistence that his driftwood sculpture *Nimis* (Latin for 'too much') be destroyed. He began constructing the sculpture in 1980, without a permit; it took the state a couple of years to notice and order him to take it down. Wooden batons have been nailed together into a swarm of tunnels, so that they resemble the shed skin of a great worm. They are narrow, tortuous, spiked with the business end of the nails that hold it together, projecting into the passageway with barbs as vicious as shark's teeth.

Safety-conscious Swedish government officials couldn't take responsibility for such a structure, and over decades have repeatedly tried to have it removed. Because it's illegal there can be no signposts, and on our first attempt to find the borders of this micronation my son and I got lost in the forest. Obliged to switch to the coast, we hopped three kilometres on giant boulders, then around a headland of cormorants – and there it was: twisting helices of driftwood between forest and beach. There were no other visitors.

Vilks died in 2021 but his legacy lives on: you can apply to become a citizen of Ladonia for free, simply by filling in some details online (choose a Latin phrase that defines you, enter the colour of your eyes, and add your height in 'toilet paper squares'). The national anthem is the sound of a stone falling into water, and its language consists of two words: 'waaaaalll' and 'ÿp'.

We scaled the second-highest of the towers, the one that seemed least likely to collapse. Eye-level with the treetops, we looked down on another family of tourists arriving at the edge of the forest. They gazed up in wonder, then, smiling wordlessly, began to climb in through the tunnels.

The centre of Gothenburg is widely acknowledged to be beautiful, but we never saw it for flyovers, cranes, its clutter of bridges, and a deluge of rain. At Tanumshede I stopped to see the Bronze Age artworks thought to be almost four thousand years old, when the sea level here was much higher and the land was home to a coastal people. Timeless scenes of sex and violence, grief and power were etched onto canvases of granite scoured clean by the Ice Age.

The old border bridge between Sweden and Norway was damaged in the Second World War. German-occupied Norway has never really forgiven Sweden for claiming neutrality while selling iron ore to the enemy – a sale that prolonged the war – and in 1942 some explosives left on the bridge were ignited not by partisans, but by lightning. The new bridge is an immense arch of reinforced concrete, almost two-thirds the size of Sydney's Harbour Bridge, built in 2005, exactly a century after Norway won its independence from Sweden. We drove swiftly over the thick white line at its centre marked 'SVERIGE-NORGE'.

My car is UK-registered, and at the frontier every Swedish vehicle was waved on but I was flagged down. The guard – white-blonde hair, white eyelashes – wanted first to know whether I had any alcohol or cigarettes, and second which part of the UK we were coming from. When I replied 'Scotland', he smiled broadly and waved me on – whether because of Scotland's Norse history, or because the country had voted against Brexit, he didn't say.

The Oslo opera house looks like a spaceship surfacing in an ice field. Behind it the Edvard Munch museum had the swishest toilets north of Copenhagen and a hundred different ways of buying a reproduction of the artist's 1893 *Skrik* (*The Scream*). A postcard was the same price there as a cup of coffee. The painting is said to be inspired by a view from a path looking over the Oslofjord, sky reddened by Krakatoa's eruptions. Munch wrote of it:

> I was walking along the road with two friends – the sun was setting – suddenly the sky turned blood red – I paused, feeling exhausted, and leaned on the fence – there was blood and tongues of fire above the blue-black fjord and the city – my friends walked on, and I stood there trembling with anxiety – and I sensed an infinite scream passing through nature.

I walked up a forest path to the viewpoint where Munch had his vision, and where a bronze plaque of the famous painting had been fixed to the parapet. The railings behind the screaming skull look identical to those of another iconic Munch painting, *The Girls on the Bridge*, painted in a small fishing village further down the coast. The following day we went there.

In Åsgårdstrand, Munch painted three young women dressed in white, red and green who stand poised between childhood and adulthood, gazing into shadowy water. I was reminded that Billy Goats Gruff is a folk tale originally from Norway, and though the painting is of girls, not goats, and there are no trolls in evidence, Munch's bridge seemed similarly emblematic of perilous crossings, with the girls suspended over a dense, unknowable and threatening darkness. 'Maybe it's the bridge between childhood and adolescence,' I said to my son. 'And the darkness is meant to represent their future.' He stood watching the water for a few moments, and I wondered if he was lost in thought, perhaps contemplating his own journey towards adulthood. Then he turned to me slowly and said, 'Do you think there's anywhere we can get something to eat?'

The Norwegian highway system is a miracle of investment infrastructure.

With its wealth Norway has built a network of bridges and tunnels that slide blithely over fjord and through mountain, skipping unobstructed along one of the more fissured coastlines of Europe. Because of its oil revenue (Norway owns the largest single sovereign wealth fund in the world), and because it separated from Sweden over a century ago, Norway has the third-highest GDP in Europe, behind only Luxembourg and Switzerland.

It was a two-hour crossing by catamaran from Kristiansand to Jutland, closing our loop of the Kattegat and sailing into the harbour that once joined Scotland to Scandinavia. Ships registered in Hirtshals would dock in Thurso on the Scottish mainland, and then again in Shetland on their way to Iceland. But with the withering of trade links, they now sail blithely past. From the Arctic coast of Scotland earlier that year I had watched one go, slipping through the Pentland Firth as if anxious it might be impounded. Seeing it pass was a small grief. 'Wait!' I wanted to shout. 'Don't leave us!'

The Norwegians and Danes on the ferry with us read copies of their respective *Dagbladet*s while Tina Turner welcomed us to the Thunderdome and the Eurythmics walked on broken glass. Our own race to get back to Amsterdam was on; unable to sail from anywhere in Scandinavia, we'd have a thousand-kilometre drive instead.

South of Århus, we made a detour to the Infinite Bridge – a circular walkway 60 metres in diameter, half onshore, half at sea. James Joyce said of piers that they are 'disappointed bridges', and it's normal for coastal communities in Denmark to have a communal boardwalk pier, for the locals to access sea-swimming without having to step barefoot over shingle and seaweed. The Infinite Bridge plays with ideas of the reconciliation of land and sea, water and air, light and sky, sea and sand, rock and cloud. It's a conduit, a passageway, a circular breath. One of its two designers, Johan Gjøde, said that nature is the true architect of the bridge; it's 'the city's skyline, the harbour and the relationship to the water that is the true art piece'.

A woman with pale skin and yellow hair emerged from between the trees, then walked naked into the water. She swam out for a few strokes before turning back to merge again with the forest. Birdsong was dense in the glade of the stream that ran directly into the halo of the bridge. A line of stones extends into the centre of the circle, the remains of an old pier for steamers.

My son and I walked out to the furthest extent of the Infinite Bridge. A swimmer crawled northwards through the sun's glitter, hauling behind him a fluorescent buoy; kayakers slid silently south, giving him a wide berth. The bridge is a bridge of the mind, but also a celebration of the Danish passion for wild swimming: as we sat, several groups came, jumped, swam, then left again, this magnificent bridge just an everyday part of their lives. Light reflected from the sea in great glamourings, a sailing boat stood at anchor, and always there was birdsong giving texture to the air.

Long before the Øresund Bridge was dreamed of, the ferry that sailed between Copenhagen and Malmö was the *Caledonia*, a ship built in Glasgow and a reminder of a time when the different nations of Britain still looked to the sea and sustained connections with their neighbours. In our nine days' journey, I'd felt keenly how much my son's Europe is different from the one I grew up in: more girdled, more anxious, its connections amputated. It feels itself under siege. For a British citizen at least there are now far fewer possibilities of welcome, of integration, of connection and cross-pollination. Who's to say that the Danes haven't benefited from shifting their focus away from the west, strengthening their links with Norway and Sweden at the expense of those with the UK. It's the island of Britain that has been impoverished.

The planet is now woven in a cat's cradle of air routes, and so perhaps planners and voters and politicians see less need for grand bridges between countries, and for maintaining their traditional sea routes. But the twenty thousand commuters a day of the Øresund link frames the question differently: the bridges of Scandinavia demonstrate what can happen when nations cooperate for mutual benefit and exchange.

Exit

Through 2022 and 2023, as the implications of Britain's departure from the European Union were becoming evident in terms of damage to the economy, politicians in Scotland began laying out the arguments for another referendum on independence. I revisited the Union Chain Bridge with my children after a gap of more than forty years. I was curious to see what their reactions would be; whether the border and the bridge had retained its power to astonish. But it wasn't possible to cross: its chains were worn out, and its long-obsolete materials needed replacement.

The site was deserted; a plaque nearby informed me that the reopening of the bridge was already many months delayed. Ropes swung from gantries as if ripped out by an angry giant; an empty cradle for the construction workers hung pointlessly from the rigging. The bridge looked forlorn, overlooked, as if in acknowledgement that all unions need maintenance and care if they're to succeed.

The bridge reopened as I completed this book, to great celebration among communities on both sides of the Tweed – Scottish and English. 'This bridge stands as a testament to partnership working,' said a councillor for the Northumberland side, Glen Sanderson, 'and shows what can be achieved when everyone is pulling in the same direction.' A representative for the Scottish Borders Council, John Greenwell, praised all the 'hard work' that had gone into restoring the bridge for a third century of use. 'It's a symbolic link between England and Scotland,' he said, 'which has now been protected for many generations to come.' In 2024 I visited again with my family to see it restored, and together with my kids strolled out to the bridge's midpoint. They jumped back and forth over the border line, Scotland to England and back again – marvelling as I had, more than four decades before, that a border could be so easily conjured, and once conjured, so easily transgressed.

As I write, archaeologists in Zambia have uncovered what might be humanity's oldest known bridge: half a million years ago some logs were whittled and fitted together by our hominid ancestors to make a kind of platform over water. Even Hadrian's bridge in Rome won't last half a million years; all bridges need work if they're to survive, and I'd seen again and again just how easily they can fall into disrepair.

Britain's proud status as an island nation with a natural border of sea is a temporary arrangement, an accident of climate. On this global journey I'd witnessed first-hand just how easily national borders too can be redrawn, and the ways connections between communities have been ceaselessly reimagined. From Peru to Singapore I'd seen legacy bridges left behind by lost empires. From Bosnia to historic Palestine I'd encountered magnificent crossings that have nevertheless been resented by the communities they were ostensibly built to serve, reminders that for a bridge to be successful, it has to have the benefit of all of its potential users in mind. From Old London Bridge to the newest Çanakkale crossing, several of the bridges I'd followed were built to bolster commerce, while from Prague to New York, others have become symbols of the ambitions of a culture. On my travels I had encountered too few bridges of peace, like Derry's crossing over the Foyle, and too many intended only to accelerate the plundering of resources – from Attock to the Zambezi. The Bridge of Freedom in Venice threatens the city it was supposed to sustain, while dam bridges like the Hume and the Yenisei make manifest the power of modern engineering to utterly transform the environment – for good and for bad. The bridges of Denmark, Norway and Sweden showed me just how much nations stand to benefit when they open themselves to free traffic and exchange. The bridges of Munch and Monet, of Turner and Joseph Stella, reminded me of something all artists know: that crossings are symbolic of transition and of transformation, and the worlds they bring us to can be heavenly or they can be hellish.

As a species we have to take better care where, and how, we build our bridges. Our reasons for constructing them are always evolving, as are our reasons for drawing up borders: the Romans' imperial frontier along the Forth didn't last for long; Scotland's with England along the Tweed might yet be hardening, while, at the time of writing, the UK's border with the Republic of Ireland looks as if it could dissolve. The line of partition in Kashmir may be becoming a little more porous, even as China's with Bhutan remains stubbornly closed.

All my life I've looked for and admired bridges of the past, and now I've begun to think more carefully about bridges of the future, actual and metaphorical, and how they might be built with more sensitivity for their users, and for the planet. There are suspension bridges now in planning stages that will float on tethered anchors, with no need for their piers to reach the sea floor. To prevent the towers of such bridges rocking with the waves, each could be trussed by giant cables to its neighbour in parallel lines, like a soprano score running high over the bassline of the sea. The Scandinavians are pioneering underwater bridges – watertight tunnels anchored to the seabed. It may even be possible to use such tunnels to create energy: as their surface moorings are thrown up and down by the waves, that motion might be transformed into electricity.

Climate engineers are, with their innovations, building bridges to a more stable future climate. There are new materials in development that will span further, more strongly, and which don't involve mass dumping of CO_2 into the atmosphere. The Çanakkale Bridge in Türkiye will not remain the world's longest span for very long; the next few decades will bring further revolutions in bridge design, making possible as-yet-unimaginable crossings.

It's a winter's night in Edinburgh and I've arranged to meet my brother for a drink after work. Though we live just a few miles apart, the Firth of Forth divides us, and our lives are often so busy that if we don't put time aside to see one another, months sometimes pass between our meetings. We have to make an effort to stay connected, but it's an effort worth making. He lives on the north side of the Forth in Fife, in what the medieval cartographers thought of as another world to Edinburgh and the Lothians – Scocia Ultramarina, or 'Scotland-beyond-the-Sea'. As I set off on my bicycle from Edinburgh, I think about all the bridges I'll cross to reach him. Because of those crossings, the journey shouldn't take much more than an hour.

From my clinic door the first bridge I'll reach will be over the Union Canal – an elegant lift-bridge from 1906. The next will be over the Water of Leith, a Victorian stone viaduct of graceful arches, built in 1861 for a railway that for many years now has stood service as a cycleway (an elegant repurposing that would have seemed wildly improbable to its engineers). The path descends towards the River Almond and over the three stone arches of Old Cramond Brig – the traditional border between the city of Edinburgh and the rural shires to its west. The piers of the bridge rest on V-shaped cutwaters that remind me of the Ponte Sant'Angelo in Rome.

The approach to the bridge sparkles with frost – these days it's only for bicycles and pedestrians. I slow to take the corner with care, and at its crown pause to watch the water – this evening with a thin patina of ice. In the slanting light of my bike's headlamp I notice something I never have in all the hundreds of times I've crossed this bridge: lettering chiselled into the stone of its parapet. The age of the bridge is revealed by the frost: 'Anno Dom 1619', says a central inscription – I knew it was old, but not *that* old. Then further along, another series of inscriptions: 'Repaired By Both Shires 1687' and 'Repaired By Both Shires 1761'. This last must have been a faulty repair, because alongside it, in slightly different script, is a message from fifteen years later: 'And Again In 1776'. Over the top of all these inscriptions, the ever-confident Victorians have etched their own commemoration: 'Repaired By Both Shires 1854'.

Nearby there's a modest plaque that has also escaped my attention until now, and reading it I learn that parts of the bridge's structure date back to the 1400s. For six centuries, then, communities on the east and west banks of this river have worked to keep their connection open. It is a revelation that the bridge has been helping people get on with their lives – to meet, trade, work and see their families – for more than half a millennium. Throughout its history it has provided work, brought convenience, and encouraged the flourishing of two communities – and for a moment I imagine all the stonemasons, horses, and rough scaffolding, all the carts of stone and mortar that over the centuries have gone into keeping it open, so that I can get home quickly and easily this night.

Twenty minutes later I'm at the Forth. The view over the estuary from the south has a cinematic quality; two road bridges and a single rail bridge stand over the water, more like gates of transcendence or emblems of the future than practical structures of steel, tar and concrete. From east to west there's the Victorian rail bridge, a leviathan of red steel, then the graceful suspension bridge I ran over for charity in 1988 (for a while the longest

bridge in the world), with its twinned bike lanes. Then there's the new, vast cable-stay bridge of steel and reinforced concrete, another record-breaker of snow-white lines. Three bridges, three styles, three centuries. I don't live in their shadow, but in their light and loft. For me they have come to stand as symbols of the ingenuity of humanity. They remind me that there are countless possibilities for human connection that are yet unexplored.

These days the Forth Road Bridge is usually deserted when I pedal across, the traffic having all moved west to the new crossing. It feels exhilarating to soar above the blackness of the water, high on my own personal skyway, effortlessly crossing what was, two thousands years ago, the Romans' northern frontier. Partway over the span I reach the point where a few desperate people each year come to end their lives but far greater numbers come to make declarations of love. *Only connect.* I stop on the empty deck by its gallery of heart-shaped padlocks, each one inscribed with the names of lovers or of loved ones, remembering Thornton Wilder's insistence in *The Bridge of San Luis Rey* that it's the connections we make that give life its meaning; that bridges are the manifestation of humanity's longing to come together. *Bridges are good to think with.* The river shimmers beneath the moon, light the language of the sky in its dialogue with the sea, and the deck beneath me glitters with frost. I take out my phone to message my brother – *Nearly there* – then push on towards the other side.

Thanks

To E., for our ongoing adventures. To Karine Polwart for the Ladybird books. To Oliver Riches for sharing stories of working on some of the finest bridges in the world, and Enrico Tubaldi for explaining his research into the ways climate change is undermining them. To Jack and Jinty Francis for giving me a house full of books to grow up in, and a strong early sense of the wonder of engineering. To Alan Francis for fraternity across the water. To Jennifer Doherty and Paige Provenzano for helping make my first experience of London so transformative. To Nicky Conway, Alasdair Reid and Stuart Waterston for saving me from myself in Prague, and to Teresa Murray and Andrea Stratilová for their local insights into Charles Bridge and the Vltava. To Daniela Ellis for Polish family hospitality. To Vivek Muthurangu, Catherine Clark and Justin Perry for their companionship in Venice and in Rome. The New York Academy of Sciences made me very welcome in Manhattan for their series on wonder in science, and Catherine Clark opened her Brooklyn brownstone to me for reconnaissance trips to Brooklyn Bridge.

In Australia: thanks to Isadora Chai, Hilda Newport, Ping Leong, Angela Webster, Donald and Philippa Johnson, and Lee Tan of Melbourne's Friends of the Earth. Without the energy, friendship and enthusiasm of Phil Stewart, and his wizardry with emergency visas, E. and I would have seen little of Pakistan. Swee Yoke Chew in Kuala Lumpur and Lesley Chan in Singapore opened their cities to us. The Tibetan staff of Delek Hospital in Dharamshala, and in particular Choetso Tsering, gave us the perfect springboard and landing ground for a trip around Ladakh and

Kashmir, and Gillian Spragens gave us a Glaswegian-Californian welcome in the latter. In Thimphu I was fortunate enough to be welcomed by both Namita Gokhale and Mita Kapur. In Sarajevo, Emir Filipović was generous with his time, knowledge and histories. Thanks to Mahmoud Muna of Kalimat Festival in Palestine, to Dina Nasser and Izzedein Hussein in Jerusalem, and Penny Johnson, Mustafa Barghouti and Raja Shehadeh in Ramallah. To Allan Little for his friendship and company in some unexpected corners of the world. To Neal Ascherson for sharing his experience of Cold War Europe, as well as his intimate knowledge of the Firth of Forth. To Alison Watt for enthusiasm and encouragement, and Laura Cumming for her expert eye (and pen). Thanks to his insightful commentary David Farrier has seen this book transform from a rough blueprint into actuality, with a stronger and I hope more elegant structure, able to bear the weight of its traffic of ideas. Iain McClure's generosity facilitated my trip to Derry and the Antrim coast. Thanks to Elena Haltrin of the Gorky Institute of World Literature in Moscow, and to the British Council for facilitating my visit to Siberia. Without the advice of Anna Henriksson I would never have discovered Ladonia. Thanks too to Marianne Mitchelson for granting me permission to use her photograph taken from the top of the Forth Bridge. I'm grateful to my Russian publishers, Eksmo, for their hospitality in Moscow, and Lorna Mackay of Invernaver for offering such a tranquil space to finish the book.

With much gratitude for the work of the Samaritans, and to the RNLI, for all they do saving lives on, above and below the water.

Thanks to Jenny Brown for the coffee and cinnamon buns as well as the contracts, and to the good ship Canongate – and in particular Francis Bickmore, Leila Cruickshank, Melissa Tombere, Brodie McKenzie, Jane Selley, Jenny Fry, Jamie Norman, Jamie Byng and Anna Frame. Thanks to Bill Johncocks, champion of the well-wrought index, for his skill and attention to detail. I owe all of you a dram.

A few passages of *The Bridge Between Worlds* have appeared in other places, in other forms, and thanks are due to their editors and producers for believing in my work and commissioning me to write for them: to Hande Zapsu Watt of the late *Istanbul Review*, which first published my reflections on the Hellespont in April 2013; to Rachael Allen of *Granta*, who published 'The Third Pole', on my experience of driving through Ladakh, in August 2013; to Thomas Jones of the London Review of Books, who published my piece 'In Thimphu' in July 2014; to Clare Walker and John Goudie of BBC Radio 3, who commissioned and broadcast my essay on the Queensferry Crossing for their series *Buildings of Britain* on 13 October 2016; and to Clare Longrigg of the *Guardian*, who published my long read 'What I have learned from my suicidal patients' in November 2019.

List of Illustrations

All images using Creative Commons (CC) licences were sourced from Wikimedia Commons. Licence terms can be found here: creativecommons.org

2 Artwork from *bridges*, copyright © Ladybird (1976). Reprinted by permission of Penguin Books Ltd.
3 Skeleton of foot, medial aspect. From Henry Gray, *Anatomy of the Human Body* (1918).
10 Royal Border Bridge (1923). Tuck DB Archive.
11 Union Chain Bridge, Berwick-on-Tweed, postcard (1942).
16 'Great Forth Run' from *Edinburgh Evening News*, p. 3 (10 May 1988). Reprinted by permission of SWNS Ltd.
17 Traffic on the Forth Road Bridge copyright © Eugene O'Brien (27 March 2017). CC 4.0.
18 Forth Road Bridge tower copyright © Gavin Francis (April 2021).
20 *The Queensferry from the South* from Sir Walter Scott, *The Antiquary* (Archibald Constable, 1816).
22 The Forth Road Bridge copyright © Mike McBey (24 June 2006). CC 2.0.
30 Old London Bridge engraving by William Henry Tom (c. 1730). Sourced from the Yale Centre for British Art.
32 London Bridge, Lake Havasu copyright © Uli Elch (22 July 1973). CC 4.0.
39 *Prague. Old bridge over the Moldau* (c. 1860). Sourced from the Library of Congress.
43 Charles Bridge illustration by N. Erichsen. From Francis Lützow, *The Story of Prague* (1902).
48 View of Castel Sant'Angelo from North by Giovanni Battista Piranesi (1748).
50 Ponte e Castello Sant'Angelo postcard (1956).
52 Le antichità Romane. Tomo IV, tav. XI. // Opere di Giovanni Battista Piranesi, Francesco Piranesi e d'altri. Firmin Didot Freres, Paris, 1835–1839. Tomo 4. By Giovanni Battista Piranesi (c. 1756).
57 Ponte della ferrovia sulla Laguna di Venezia, unknown author (1910).
58 *Rialto Bridge* by Carlo Naya (1875).

61 Bridge of Sighs, Venice (1906). Tuck DB Archive.
63 The Fra Mauro map by Fra Mauro (1459).
66 Aerial photograph of Victoria Falls (1924). Tuck DB Archive.
67 Railway bridge, Victoria Falls (c. 1924). Tuck DB Archive.
74 View of Manhattan from Brooklyn by Irving Underhill (c. 1913). Sourced from the Library of Congress.
75 'Brooklyn Bridge Late Afternoon', *Post Cards: New York Series 1* by Rachael Robinson Elmer (1916). Reba and Dave Williams Collection, Gift of Reba and Dave Williams. National Gallery of Art.
77 George Washington Bridge and Hudson River, New York, postcard (1930).
79 Brooklyn Bridge copyright © Gavin Francis (February 2020).
81 Forth Rail Bridge copyright © Gavin Francis (March 2018).
82 View from cantilever of the Forth Rail Bridge by Marianne Mitchelson (April 2017). Used by kind permission of Marianne Mitchelson.
83 Alternative designs for the Forth Bridge by Wilhelm Westhofen (1890).
84 *Landscape with the Fall of Icarus* by Pieter Bruegel (c. 1555).
86 'Sudden Shower Over Shin-Ohashi Bridge and Atake (Ohashi Atake no Yudachi)', *One Hundred Famous Views of Edo* by Utagawa Hiroshige (1857).
90 Golden Gate Bridge, San Francisco, postcard by Stanley A. Piltz (1939).
94 Golden Gate Bridge image still from *Vertigo* trailer (1958).
96 Golden Gate Bridge copyright © Almonroth (16 February 2011). CC 3.0.
99 Scottie and Midge image still from *Vertigo* trailer (1958).
102 Machu Picchu illustration from Sir Clements Robert Markham, *The Incas of Peru* (1912).
110 Bosphorus Bridge, Istanbul, Turkey copyright © JoopAnt (1973). CC 4.0.
114 Smyrna, caravan bridge (1860). Sourced from the Library of Congress.
115 'Phoenician sailors building a pontoon bridge across the Hellespont for Xerxes I of Persia enabling him to invade Greece, c. 480 BC' by A.C. Weatherstone (1915).
117 Çanakkale Bridge copyright © Zafer (27 March 2022). CC 4.0.
122 Attock Bridge postcard, author unknown (1890s).
124 Fortified Northwestern Railway bridge over the Indus at Attock by William Henry Jackson (1895). Sourced from the Library of Congress.
130 Tributaries of the Indus, Ladakh copyright © Gavin Francis (September 2007).
137 Descending towards the Vale of Kashmir copyright © Gavin Francis (September 2007).
144 Anji Bridge, China copyright © wanghongliu (16 March 2020). CC 1.0.
145 Anji Bridge, China copyright © Siyuwj (6 September 2020). CC 4.0.
148 North Boat Quay, Singapore, postcard (c. 1900).
150 Cavenagh Bridge, Singapore copyright © William Cho (24 March 2007). CC 2.0.
155 Sydney Harbour bridge souvenir postcard, 1932.
157 Hume Dam from above, copyright © Gavin Francis (March 2008).

159 Hume Dam, Australia (1954). State Library of Victoria.
167 Deck of North Tower under construction, Queensferry Crossing copyright © Gavin Francis (September 2016).
168 Central and Southern Towers, Queensferry Crossing copyright © Gavin Francis (September 2016).
170 Central Tower and deck, Queensferry Crossing copyright © Gavin Francis (May 2016).
174 Map of China and India. Published by the Central Intelligence Agency (2006). Sourced from the Library of Congress.
176 Wangdue Zam bridge by Samuel Davis (1783).
181 Poster for the Trans-Siberian Railway in the 1900 Exposition Universelle of Paris (1900).
182 Railway bridge over Yenisei in Krasnoyarsk, Russia (1904). Sourced from the Library of Congress.
184 The Krasnoyarsk Dam in Russia copyright © Alex Polezhaev (2010). CC 2.0.
191 Royal Engineers Officers at the official opening of the Allenby Bridge (1918).
194 New Allenby Bridge over Jordan (1934).
197 Jerusalem seen from the Mount of Olives (c. 1905). Tuck DB Archive.
203 'A view on the Jordan' postcard (date unknown).
209 Mostar Bridge in Bosnia and Herzegovina copyright © Gavin Francis (April 2019).
211 Latin Bridge, Sarajevo, unknown artist (c. 1910s).
215 The Mehmed Paša Sokolović Bridge in Višegrad (c. 1890s). Sourced from the Library of Congress.
224 View of Derry City & Peace Bridge from the Ebrington Barracks by Gavan Connolly (14 January 2012).
226 The Tavanasa Bridge in Switzerland (c. 1907).
228 Salginatobel Bridge in Switzerland copyright © Gavin Francis (April 2021).
229 Salginatobel Bridge passageway copyright © Gavin Francis (April 2021).
232 The Little Belt Bridge in Denmark copyright © Heb (27 December 2011). CC 3.0.
233 The Great Belt Bridge in Denmark copyright © Lonni Besançon (12 April 2018). CC 2.0.
235 Øresund Bridge in Denmark and Switzerland copyright © Håkan Dahlström (9 February 2014). CC 2.0.
238 Svinesundsbrua on the border between Sweden and Norway copyright © Håkon Aurlien (14 October 2005). CC 3.0.
239 *The Girls on the Bridge* by Edvard Munch (1899).
241 Infinite Bridge in Denmark copyright © Gavin Francis (July 2022).
244 Union Chain Bridge copyright © Gavin Francis (April 2024).
248 The three Forth Bridges copyright © Roddy McDowell (March 2019). Used by kind permission of Roddy McDowell.
249 Forth Road Bridge and the Queensferry Crossing copyright © Gavin Francis (December 2023).

Selected Reading

I have read scores of books about the meaning and history of bridges during the writing of this book, but the following general titles have been among the most precious, and all come highly recommended.

Bridges of the World, by Giancarlo Ascari and Pia Valentinis, translated by Katherine Gregor (OH Editions, London, 2022)

The Forth Railway Bridge: A Celebration, by Anthony Murray (Mainstream, Edinburgh, 1983)

Bridges: A History of the World's Most Spectacular Spans, by Judith Dupré (Black Dog, New York, 2017)

bridges, by Robert Loxley, illustrated by Gerald Witcomb and Gavin Rowe (Ladybird Books, Loughborough, 1976)

Billy Goats Gruff, by Fran Hunia, illustrated by John Dyke (Ladybird Books, Loughborough, 1977)

Of Bridges: A Poetic and Philosophical Account, by Thomas Harrison (University of Chicago Press, Chicago, 2021)

Notes on Sources

9 'They do nothing for us . . .' Quoted in Katie Dawson, 'Berwick-upon-Tweed: English or Scottish?', BBC News, 1 May 2010

12 'As I went along . . .' Samuel Smiles, *The Life of Thomas Telford* (Cambridge University Press, 2014)

13 'An unofficial survey . . .' Katie Dawson, 'Berwick-upon-Tweed: English or Scottish?', BBC News, 1 May 2010

13 'A Bridge of wondrous length . . .' John Milton, *Paradise Lost* (Penguin Classics, 2003)

15 'Scotland is not content . . .' *The Queen Opens Forth Bridge*, Pathé Films (1964)

21 'Well! we shall be pretty comfortable . . .' Sir Walter Scott, *The Antiquary* (Oxford World Classics, 2009)

21 'There it stands . . .' Robert Louis Stevenson, *Kidnapped* (Penguin Classics, 2007)

22 'What is lovable in man . . .' Friedrich Nietzsche, *Thus Spake Zarathustra*, translated by Thomas Common (Gutenberg, 1999)

27 'The old river in its broad . . .' Joseph Conrad, *Heart of Darkness and Other Tales* (Oxford World Classics, 2008)

28 'Traffic of the great city . . .' Conrad, *ibid.*

30 'In the end its engineers . . .' Hubert Shirley-Smith et al., 'The Middle Ages', The history of bridge design, Britannica

31 'As fine an example . . .' *Encyclopaedia Britannica* 11th edition (Cambridge University Press, 1910–11)

34 'I listened to interviews online . . .' The Thinking Mind Podcast: Psychiatry & Psychotherapy, 'Interview #9: Building Bridges in Intercultural Therapy (with Baffour Ababio), July 2020

34 'Bridging the different parts . . .' Rosemary Gordon, *Bridges: Metaphor for Psychic Processes* (Routledge, 1993)

34 'the sword that . . .' Angela Carter, 'Adventures at the End of Time', *London Review of Books*, 7 March 1991
35 'Ships, towers, domes, theatres . . .' William Wordsworth, *Selected Poems* (Penguin Classics, 2004)
36 'He made a human figure . . . Berthold Auerbach, *Spinoza: A Novel*, translated by E. Nicholas Leipzig (Bernhard Tauchnitz, 1882)
40 'There's a story I was told about Bohemia's place . . .' With thanks to Neal Ascherson
41 'All three of them now . . .' Franz Kafka, *The Trial*, translated by David Wyllie (Gutenberg, 2005)
41 'He spied between the railings . . .' Franz Kafka, *The Judgement*, translated by Schocken Books, 1948 (Doubleday, 2009)
42 'wanted to see the Vlatava . . .' Milan Kundera, *The Unbearable Lightness of Being*, translated by Michael Henry Heim (Faber, 2000)
46 'The first rule is . . .' Marcus Aurelius, *Meditations*, translated by Maxwell Staniforth (Penguin Classics, 1964)
49 'So rooted in the Roman mind . . .' Quoted in Thomas Harrison, *Of Bridges* (University of Chicago Press, 2021)
53 'All of this was packed . . .' Dante Alighieri, *The Inferno of Dante Alighieri*, A Rhymed Translation by Seth Zimmerman, copyright © 2003 Seth Zimmerman and Janet Van Fleet
55 'And let no man marvel . . .' Marco Polo, *The Million*, quoted in Noah Brooks, *The Story of Marco Polo* (Century, 1920)
57 'In Venice . . .' Jan Morris, *Venice* (Faber, 1960)
59 'Bridges like giant gymnasts . . .' F.T. Marinetti, *The Founding and Manifesto of Futurism* (multilingual edition) (Art Press Books, 2016)
59 'All towns are the same . . .' Robert Byron, *The Road to Oxiana* (Vintage, 2010)
62 'As it is shown . . .' P. Falchetta, Fra Mauro's World Map (Brepols N.V., 2006)
69 'I always believed . . .' Saul Bellow, *The Adventures of Augie March* (Penguin, 2001)
71 'Cross from shore . . .' Walt Whitman, 'Crossing Brooklyn Ferry' from *Leaves of Grass* (Penguin Classics, 2017)
71 'Face to face for the last time . . .' F. Scott Fitzgerald, *The Great Gatsby* (Penguin, 2000)
73 'O harp and altar . . .' Hart Crane, *The Bridge* (Black Sun Press, 1930)

76 'always the city . . .' F. Scott Fitzgerald, *The Great Gatsby* (Penguin Classics, 2000)

76 'When your car moves up the ramp . . .' Le Corbusier, *When the Cathedrals Were White* (McGraw-Hill, 1964)

77 'And you that shall cross . . .' Walt Whitman, 'Crossing Brooklyn Ferry', from *Leaves of Grass* (Penguin Classics, 2017)

78 'There now is your . . .' Herman Melville, *Moby-Dick* (Penguin Classics, 2003)

78 'Unsettled possession . . .' Henry James, *The American Scene* (Penguin Classics, 1994)

78n 'Bequeath to us . . .' Hart Crane, 'Voyages II' from White Buildings: Poems (Boni & Liveright, 1926)

79 'Climactic ornament . . .' Marianne Moore, 'Granite and Steel', *The New Yorker*, 1966

80 'One feature especially delights . . .' Quoted in Iain Boyd White and Angus J. Macdonald, *John Fowler, Benjamin Baker, Forth Bridge* (Edition Axel Menges, 1997)

81 'Embodied pure functional utility . . .' D'arcy Wentworth Thompson, *On Growth and Form* (Cambridge University Press, 1917)

81 'the supremest specimen . . .' William Morris, at the Art Congress of Edinburgh (October 1889), reported in *Supplement to the Railway News*, 8 March 1890

81 'It is, of course . . .' Sir Benjamin Baker, quoted in Thomas Mackay, *The Life of Sir John Fowler, Engineer* (Cambridge University Press, 2013)

82 'The broad river . . .' Wilhelm Westhofen, *The Forth Bridge* (Offices of 'Engineering', 1890)

85 'how everything turns away . . .' W. H. Auden, 'Musée des Beaux Arts', from *Collected Poems* (Vintage, 1991)

85 'the whole pageantry . . .' William Carlos Williams, 'Landscape with the Fall of Icarus', from *Collected Poems: 1939–1962* (New Directions Publishing, 1988)

89 'The Golden Gate Bridge is . . .' Quoted in Joan Libman, 'Golden Gate Bridge: Triumph, Tragedy: Suicide Rate Shadows the Span's 50th-Anniversary Celebration', *Los Angeles Times*, 22 May 1987]

89 *The Bridge*, directed by Eric Steel (2006)

90 'As everyone knows . . .' Simon Critchley, *Notes on Suicide* (Fitzcarraldo, 2020)

95 'Macadam, gun grey . . .' Hart Crane, *The Bridge* (Black Sun Press, 1930)

98 'Vertigo is a film . . .' Rebecca Solnit, *A Field Guide to Getting Lost* (Canongate, 2006)

100 'It was destroyed . . .' Clements Markham, *The Incas of Peru* (Dutton, 1912)

100 'There's a film made in the 1930s . . .' *The Adventure Parade*, Castle Films 'African Pygmy [sic] Thrills', uploaded by Geo. Willeman, YouTube, 20 Sept 2011

103 'I stepped from Plank . . .' Emily Dickinson, 'I stepped from Plank to Plank', from *The Selected Poems of Emily Dickinson* (Rock Point, 2022)

105 'The sanctuary was lost . . .' Hiram Bingham, *Lost City of the Incas* (Weidenfeld & Nicolson, 2003)

106 'Soon we shall die . . .' Thornton Wilder, *The Bridge of San Luis Rey* (Penguin Modern Classics, 2000)

108 'The Hellenic race . . .' Aristotle, *Politics*, translated by Benjamin Jowett (Clarendon, 1885)

112 'I could not name . . .' Homer, *The Iliad*, translated by A.T. Murray (William Heinemann, Ltd., 1924)

114 'To abduct women . . .' Herodotus, *Histories*, quoted in Roberto Calasso, *The Marriage of Cadmus and Harmony* (Vintage, 1994)

115 'Xerxes made libation . . .' Herodotus, *The Histories*, translated by George Campbell Macauley (Macmillan, 1890)

119 'On the other side . . .' Mountstuart Elphinstone, *An Account of the Kingdom of Caubul* (John Murray, 1819)

120 'The ferocity of the Zulu . . .' W.S. Churchill, quoted in Kirk R. Emmert, *Winston S. Churchill on Empire* (Carolina Academic Press, 1989)

123 'I have often wondered . . .' *The Baburnama*, translated by W.M. Thackston Jr. (Modern Library Inc., 2002)

123 'Truly the Grand Trunk Road . . .' Rudyard Kipling, *Kim* (Vintage Classics, 2010)

124 'It must be manifest to you . . .' Mohandas K. Gandhi, *The Penguin Gandhi Reader* (Penguin, 1995)

127 'There can be no question . . .' Lord Mountbatten, *Broadcast by Viceroy*, quoted in *The Times*, 4 June 1947

141 'Under heaven nothing is . . .' *Tao Te Ching*, translated by James Legge (Clarendon, 1890)

145 'This stone bridge . . .' Joseph Needham, *The Shorter Science and Civilisation in China*, Vol. 5 (Cambridge University Press, 1978)

147 'The fallout of the assault . . .' António Guterres, UN Secretary General, 'Address to Columbia University', 2 December 2020

148 ' . . . all the charming chaps . . .' Martha Gellhorn, *Travels with Myself and Another* (Eland, 2002)

152 'without a drastic reduction in CO_2 emissions . . .' 'Anticipating Future Sea Levels', Nasa Earth Observatory, 7 July 2021: https://earthobservatory.nasa.gov/images/148494/anticipating-future-sea-levels

152 'We live in a world where . . .' Quoted in Elizabeth Kolbert, *Under a White Sky* (Vintage, 2022)

154 'Mr Speaker, I move that . . .' Kevin Rudd, quoted in Catriona Elder, *Dreams and Nightmares of a White Australia* (Peter Lang, 2009)

157 'In 1837 a London *Report* . . .' *Report of the Parliamentary Select Committee on Aboriginal Tribes* (British settlements), 20 February 1837: https://apo.org.au/ node/61306

158 'If the workforce of a colony . . .' Henry C. Morris, *The History of Colonisation* (Macmillan, 1900)

158 'The survival of the natives . . .' George Henry Lane-Fox Pitt-Rivers, *The Clash of Cultures and the Contact of Races* (Routledge, 1927)

158 'In 1936 . . .' For this chronology I'm indebted to Sven Lindqvist's book about dispossession and the genocidal attitudes of successive Australian administrations, *Terra Nullius* (Granta, 2007)

158 'It was we who . . .' Quoted in Stuart Macintyre, *A Concise History of Australia* (Cambridge University Press, 2009)

159 'We apologise for the laws . . .' Quoted in Catriona Elder, *Dreams and Nightmares of a White Australia* (Peter Lang, 2009)

165 'She erected dwellings . . .' Bishop Turgot, *Life of St Margaret, Queen of Scotland*, translated by William Forbes-Leith (William Paterson, 1884)

172 'In perfect balance . . .' *Dhammapada*, 10:144, translated by Frank Lee Woodward (Madras, 1921)

173 'This same Orient . . .' Roland Barthes, *Mythologies*, translated by Annette Lavers (Vintage, 1993)

178 'Bare banks, canals without quays . . .' Alexander Pushkin, *Prose Stories*, translated by T. Keane (G. Bell & Sons, 1916)

180 'When the writer Colin Thubron . . .' Colin Thubron, *In Siberia* (Vintage, 2009)

190 'The importance of Jerusalem . . .' Quoted in Eric H. Cline, *Jerusalem Besieged* (University of Michigan Press, 2004)

191 'How was this piece of dark wood . . .' Mourid Barghouti, *I Saw Ramallah*, translated by Ahdaf Soueif (Bloomsbury, 2005)
191 'Then Moses went up . . .' Deuteronomy 34:1–3
192 'So Moses the servant of the Lord . . .' Deuteronomy 34:5
192 'And the Gileadites . . .' Judges 12:5–6
193 'The Jordanians call it the . . .' Mourid Barghouti, *I Saw Ramallah*, translated by Ahdaf Soueif (Bloomsbury, 2005)
195 'Sebald's literary aim . . .' 'Changing perspectives on W. G. Sebald, 5 minutes with Uwe Schütte', Liverpool University Press blog, 30 August 2018
195 'The largest fortifications . . .' W. G. Sebald, *Austerlitz* (Penguin, 2011)
195 'Israel then had the highest military spending . . .' ($2,508) Bastian Herre et al., 'Military Personnel and Spending', Our World in Data: https://ourworldindata.org/military-spending
199 'The anatomy lessons . . .' W. G. Sebald, *The Rings of Saturn* (Vintage, 2019)
200 'Israel is not a state . . .' Quoted in Elsa Auerbach et al., 'The Safety of Others', *London Review of Books*, 5 April 2022
200 'The hereditary custodians at the gate . . .' Quoted in Alan Balfour, *The Walls of Jerusalem* (Wiley, 2019)
205 'In a war in which . . .' Laura Silber & Allan Little, *The Death of Yugoslavia* (Penguin/BBC Books, 1995) Used by kind permission of Allan Little
209 'Stone collar . . .' Ismail Kadare, *The Three-Arched Bridge*, translated by John Hodgson (Vintage, 2013)
213 'Even the least of the townsmen . . .' Ivo Andrić, *The Bridge Over the Drina*, translated by Lovett F. Edwards (Harvill, 1994)
215 'Thus the generations . . .' Ivo Andrić, *The Bridge over the Drina* (Harvill Press, 1994)
219 'The EU threatens . . .' Speech by John Hume, 'Europe as a force for creative reconciliation', University of Ulster at Magee, 26 April 2004: https://cain. ulster.ac.uk/cain/john_hume. Used by kind permission of The John & Pat Hume Foundation
220 'If he's got £20 billion . . .' Nicola Sturgeon, quoted in 'Work "under way" into Scotland–Northern Ireland bridge feasibility', BBC News, 10 February 2020
221 'Oh my God, don't forget . . .' Lisa McGee, *Derry Girls*, Hat Trick Productions, first broadcast 5 March 2019
222 'In the winter of 1688 . . .' Susan McKay, 'In Portadown', *London Review of Books*, 10 March 2022

223 'How can the symbols . . .' Lewis Hyde, *A Primer for Forgetting* (Canongate, 2019)

224 'Even in the darkest moments . . .' Quoted in 'John Hume: Funeral hears that Nobel laureate "never lost faith in peace"' BBC News, 5 August 2020

225 'Reinforced concrete does not . . .' Robert Maillart, quoted in Judith Du Pré, *Bridges* (Black Dog, 2017)

230 'The Idea: a circle . . .' engraved on a bilingual plaque at the site of The Infinite Bridge

238 'I was walking along the road . . .' Peter Russell, *The Delphi Complete Paintings of Edvard Munch* (Delphi, 2017)

240 'the city's skyline . . .' Gjøde, Johan, "*The Infinite Bridge / Gjøde & Povlsgaard Arkitekter*" 14 Jul 2015. *ArchDaily*. Accessed 9 Apr 2024

243 'This bridge stands as . . .' Giancarlo Rinaldi, 'Union Chain Bridge linking Scotland and England reopens', BBC News, 17 April 2023

Index

Italic page references indicate illustrations; the suffix 'n' a footnote.